The Best American Science and Nature Writing 2000

The Best American Science and Nature Writing 2000

Edited and with an Introduction
by David Quammen

Burkhard Bilger, Series Editor

HOUGHTON MIFFLIN COMPANY
BOSTON · NEW YORK 2000

ISSN 1530-1508
ISBN 0-618-08294-8
ISBN 0-618-08295-6 (pbk.)

Printed in the United States of America

QUM 10 9 8 7 6 5 4 3 2 1

Contents

There are limits, granted. Does our definition of writing include reports in scientific journals? Poetry? Prose poems? No, no, and no, though some passages by Peter Matthiessen and Anne Fadiman are poetic enough. Does straight reporting count? Yes, we decided, so long as the style is literary and its purpose broader than news gathering. Book excerpts are fine, too, but only if they appeared previously in a magazine and are truly self-contained. (Natalie Angier's essay on evolutionary psychology, taken from *Woman: An Intimate Geography,* qualifies on both counts.) But novels, commencement addresses, cartoons, and plays — even a Tom Stoppard play on the second law of thermodynamics — fall outside our purview.

That covers the basics, but it leaves the thorniest questions unanswered. How broadly do we define *science,* for instance? Until a year or two ago, a science magazine like *Discover* rarely published stories on medicine or technology, calling these fields applied science rather than science proper. But that standard seems more arbitrary every year. Quantum physicists have colonized Wall Street and microbiologists have defected to the biotech industry in droves; mathematicians are programming computer games and chemists are creating laundry detergents. Some of the best science stories cover research where you least expect it: in camel racing ("Lulu, Queen of the Camels"), for instance, or in Mormonism ("This Is Not the Place").

As you might think, such exotic birds rarely fly in flocks. You'll find a few in *The Sciences, Scientific American,* and *American Scientist,* but science writing, in the main, is still a didactic genre. The classic feature format, perfected by an earlier incarnation of *Scientific American,* starts with a few mildly diverting sentences and then gets down to business: page after page of explanation, relieved only by the occasional chart or graph. Most of the time that's all for the best — who wants storytelling when you're trying to understand particle physics? — but it leaves slim pickings for anthologists. Even science bestsellers like *A Brief History of Time* tend to be admired more for their lucidity than for their literary daring.

Nature writing, as David Quammen notes in his introduction, often suffers from the opposite tendency. As a result, most of these pieces were found in general-interest magazines of the literary sort — places where science and nature are treated as just another subject for writers to bring to life. Still, some of the most distinctive

Foreword

I'VE NEVER BEEN bird-watching, but after months of searching out these stories in the New York Public Library, of hiking up marble canyons and through stacks of compacted trees, I know how it must feel. One day you see a flash of beguiling color — a lovely opening paragraph, say, or a compelling thesis — only to lose it in a thicket of confusing prose. The next day you stare at something for a moment and dismiss it as ordinary, only to catch your breath when the sun strikes its wings. You might spend hours tracking a familiar singer — be it Andrea Barrett or E. O. Wilson — through card catalog and database, across the mountains of Lexis-Nexis and into the valley of ProQuest Direct, only to find that her or his song hasn't been heard all year. There is no lack of birds, of course, but most are sparrows and grackles, and you're after something rarer and not quite so noisy.

The problem, first of all, is deciding what to seek and where to seek it. Great science and nature stories don't come precategorized in official lists. They don't cleave to a single, recognizable form. Their one common trait is longevity — no matter how timely or rich in specific detail, the pieces that follow should still be worth reading in five or ten years, if not longer — but they shouldn't sacrifice immediacy for timelessness, information for reflection. This book is devoted to the best American science and nature *writing*, David Quammen points out, not the best American science and nature *essays*. For better or worse, it comes with a wide-angle lens, and so dooms us to more than a few wild-goose chases.

voices come from smaller, more secluded places. Ken Lamberton's essay on toads — as vivid and affecting as it is unexpected — was written in prison and published in *Puerto del Sol.* Wendy Johnson's meditation on death and gardening comes from *Tricycle,* the Buddhist review. Paul De Palma's incisive critique of the popular obsession with computers appeared in the *American Scholar.*

Ironically, in this context, De Palma's piece will raise a question exactly opposite to the one he intended: Why nothing from the Internet? E-mail has made writers — or at least typists — of us all, and the on-line landscape is dotted with great piles of science and nature writing. Is nothing worth saving in all those virtual haystacks? Well, yeah, probably. But searching them might take a lifetime and find hardly a needle. Even the best Internet magazines (*Slate, Salon*) tend to publish articles that are either too chatty or too news-oriented — too mindful of our impatience with reading from a screen — to hold up in a collection like this one.

How widely that approach will spread to print remains to be seen. For now, the country still has hundreds of literary journals and magazines willing to publish lengthy, provocative work on a stunning range of topics. But less and less of it seems to sink in. When I asked various editors, writers, and journalism professors to suggest stories for this volume, they invariably came up blank. The reason isn't that science and nature writing has been less than memorable this year — these pieces bear witness to that — but that our minds have been bombarded into impermeability. Like long-time New Yorkers, who walk past the most poignant street dramas without blinking, we've grown so adept at filtering information that we sometimes miss what's most important.

The purpose of this book, then, is not only to celebrate, delight, and inform but also to remember and preserve. As Alexander Stille wrote in *The New Yorker* last year, in an article on the alarming accumulation and deterioration of digital archives in Washington, "The danger is not that some modern Sophocles will be totally lost . . . but, rather, that such a vast accumulation of records makes it nearly impossible to distinguish the essential from the ephemeral." This series, we hope, will offer future readers one guide to the essential.

It has been a pleasure and an honor to work with David Quammen on this volume, after having admired his writing for so long. I would have loved to include one of his pieces among our selections, but his advice, suggestions, and eloquent introduction more than make up for the loss.

I want to thank Laura van Dam, my editor at Houghton Mifflin, for roping me into this project and for corralling it to completion with such grace and skill. Robert Atwan, the creator and series editor of *The Best American Essays,* suggested several stories and pointed me to dozens of wonderful journals. My friend Todd Wiener helped put together a database of nearly two hundred editors, prepared a mail merge, and showed me the fastest way to stuff envelopes. Finally, my love and gratitude go to my wife, Jennifer Nelson, and my children, Hans and Ruby, for putting up with all those sojourns to the library, and for welcoming me back with open arms.

Submissions for next year's volume should be sent, with a very brief cover letter, to Burkhard Bilger, c/o Editor, The Best American Science and Nature Writing 2001, Houghton Mifflin Company, 222 Berkeley Street, Boston, MA 02116.

BURKHARD BILGER

Introduction: The Vine-Tree

SCIENCE, like democracy and tai chi and golf, is a human activity. It's not a body of Truth, inherent to the universe and revealed by priests and priestesses in white lab coats. It's not irrefragable, nor even so purely objective as it sometimes pretends. Science is a subset of human culture, which is a subset of primate behavior, which in turn is a subset of nature. That's partly why, beyond merely being important, it's so damned interesting. People do science just as people do marriage or baseball — sometimes successfully, sometimes gracefully, sometimes badly. But in the moment of history in which we presently live, those nested relations — science within culture within nature — can easily be forgotten, and the hierarchy of scale can seem reversed. To say that in plainer words: Science looks big nowadays, and nature (as it is carelessly, narrowly, too often construed) looks small. Furthermore, science is getting ever bigger and more potent, whereas nature in the narrow sense is getting smaller, piece by piece, like a pizza on a platter between teenage boys. It's shrinking away, as animal species, plant species, whole ecosystems disappear down humanity's hungry maw.

Setting aside (for now) the dire subject of that shrinkage, that ruinous drawdown of biological diversity on Earth, let's broaden our thoughts by construing "nature" more carefully and inclusively. A volcanic eruption on Mars is nature. A black hole is nature. The atomic reactions occurring within the stars Mizar and Alkaid are nature. Dark matter is nature, as are protons, neutrons, iron, manganese, bismuth, osmium, and iridium (but iridium.com is definitely culture). Chalcedony and cinnabar are nature; so too are goethite and berzelianite and samarskite and bunsenite (some of Oliver Sacks's favorite minerals during his chemistry-obsessed boyhood, as he recollects in this volume), though the names by which we know them are cultural. Chemistry itself is a science and therefore a cultural construct, as is the periodic table, even while the relations it graphically displays are part of nature. String theory is culture, but the strings themselves, those infinitesimal, twanging, hypothetical filaments suspected to be the tiniest components of the universe (they can be "closed loops like rubber bands or open-ended like bits of twine," according to Gary Taubes's report here), are nature — if they exist at all. Gravitinos and quarks may be nature or figments of speculation, I don't feel qualified to say. The fallen meteorite ALH 84001 is nature, whether or not it contains traces of extraterrestrial bacteria. The big red spot on Jupiter, garish, inscrutable, is nature. This is a more cheerful as well as a more capacious view of what nature encompasses — don't you agree? — since it puts the whole thing nearly beyond the scope of annihilation by wee, puissant us. Humankind, with its global mastery and its limitless appetites, seems destined to eat up most of Earth's biological diversity and poop it out in the form of plywood, fish sticks, beer cans, and silicon chips, but at least there will still be hydrogen atoms fusing to form helium at the cores of distant stars and an occasional comet whomping into the surface of planet Zork. Astrophysics is culture, and astrophysics is big, but of nature there's an entire universe.

One advantage to this broader definition is that it allows for a rich range of topics within the book you now hold, and within the others in the annual series that will follow it. By seeing nature big, seeing it as the realm within which human enterprises such as science, religion, and historiography occur, we may free ourselves from most of the old expectations and biases, positive and negative, associated with what has commonly been understood by the term "nature writing." Personally I favor that dissociation, because my own biases are largely negative. Although nature writing in the traditional sense does include some clarion acts of literature (by Thoreau, Muir, and Leopold, for instance) and many other fine, steady, attentive works of observation and reflection, there has also been quite a lot of ethereal tripe. There have been too many purple effusions by one writer or another standing out in a forest or a meadow, misted by rain or spring pollen or pheromones and transfixed by the spectacle of his or her own sensitivity. Too high a tolerance for such stuff, too great a willingness to let the merits of nature as subject excuse pious sentimentalism and self-absorption, have tended to ghettoize nature writing in our time. People who love it read it; most other readers, especially those of more urban and trenchant disposition, hear the first chirp of a lark, the first babble of a mountain freshet, and run the other way.

That's an unfortunate bifurcation with possible consequences. For the good of all concerned, we need the urban and trenchantly disposed to remain receptive to the subject of nature, in both its broad physical sense and its narrow biological sense; and we need the self-identified lovers of nature to maintain an edge of critical intelligence. We can't afford to take nature's command of attention, of respect, for granted. On the other hand, we can't afford to let nature seem optional, a fancy best left to the fanciers. Nature is total and elemental. It's *the world* — in the big sense of *world*, meaning universe, not Earth — and though we can damage our own little bit of that universe, degrade it, render it boring and ugly, we can't ever escape the larger context. Diversity disappears, but not nature. Extinction is nature. A planet once green and acrawl with living variousness, latterly paved with concrete and stacked with tall buildings full of humans, humans, and only more humans — well, that's nature too.

Granted, no piece of writing pleases everybody. But if the mix of ideas and styles and tones and attitudes and topics within a given body of literature is raucously heterogeneous, that literature will be more likely to reach beyond the ghetto. Such a mix is the guiding ideal of this book.

But what about science writing, as commonly understood? Is it a tradition separate from nature writing, and has that always been so? No, the two have long been connected, sometimes closely and sometimes less so. But as science has defined itself more clearly (over the past several centuries) and asserted itself more forcefully (especially within the last six decades), the trend has generally been toward autonomy for its literature also.

In the English language, the tradition of science-and-nature writing for a general readership can be traced back along two distinct lineages. The first takes us to Gilbert White, an obscure village curate, whose book *The Natural History of Selborne* was published in 1788 and later became one of the most permanently cherished works of English literature. (According to a recent tally, it stands fourth on the overall list of reprint editions.) White wrote of the fauna, flora, and seasonal rhythms in his own little village in eastern Hampshire, and his legacy can be thought of as the Stay Home and Observe with a Gentle Heart school of science-and-nature writing. In France, J. Henri Fabre became the headmaster of that school.

The other lineage also has its origins in the eighteenth century and before, but it is best represented by Charles Darwin, who published his not-so-catchily titled first book, *Journal of Researches into the Geology and Natural History of the Various Countries Visited by H.M.S. Beagle, Under the Command of Captain Fitzroy, R.N. from 1832 to 1836*, three years after he returned from that round-the-world journey. In later editions and reprints, the book became better and more conveniently known as *The Voyage of the Beagle*. There had been earlier instances of a ship's naturalist publishing a book based on his notes from a sea journey of exploration, notably *Observations Made During a Voyage Round the World . . .* , et cetera, another prolixly titled but interesting record, by one Johann Reinhold Forster, a prickly Prussian who sailed with James Cook. But unlike the others, Darwin's became a bestseller, making him a famous young naturalist and setting the stage for him to startle a large rather than a small audience with *The Origin of Species* twenty years later. Darwin's *Journal* became paradigmatic for the Go Forth and Observe with a Probing Mind school, to which Joseph Hooker, T. H. Huxley, Henry Walter Bates, Henry O. Forbes, Alfred Russel Wallace, and others contributed during the Victorian era. (I'd add Mary Kingsley, whose journeys up the Ogooué and Congo rivers were as adventurous as any of those others, but she was more of a travel writer, an observer of people and customs, than a literary naturalist.) On the American side of the Atlantic, William Bartram had worked in the Go Forth manner, publishing his *Travels* in 1791 after having groped through backcountry Florida, Georgia, and the Carolinas at the time of the Revolutionary War. Thoreau himself did a little bit of Go Forth (*The Maine Woods* and *A Week on the Con-*

cord and Merrimack Rivers) and a lifetime of Stay Home (as reflected in his fourteen-volume journal). *Walden* itself was a Stay Home enterprise masquerading as a Go Forth. He subtitled it *Life in the Woods*, but he walked into Concord frequently during that pondside sojourn to refresh himself from his mother's cookie jar.

The Stay Home/Gentle Heart approach was carried forward in America by some elegant and distinctive writers, including Henry Beston, Joseph Wood Krutch, Aldo Leopold himself (at least after focusing on Sand County), E. B. White (again, after retiring from Forty-third Street in New York City to Maine), and more recently, Wendell Berry. The Go Forth/Probing Mind approach has continued among that blessed group of scientists who, like Darwin and Wallace, have been able to describe their efforts, observations, and ideas in engaging prose — a group that in our own time is well represented by George Schaller, Jane Goodall, Ernst Mayr, Edward O. Wilson, Richard Nelson, and Bernd Heinrich.

At this point I'll make my apology that of course this binary schema is crude and ridiculously inadequate. The particularities of all those decades of writing are much more intricately woven, like the various forking, wrapping, and converging shoots of a strangler fig. As the fig plant thickens into maturity, one line of growth merges against or overlaps another. For instance, some of the most probing minds (I think of Wendell Berry) have mainly stayed home, and some of the gentlest hearts (I think of Jane Goodall) have gone forth into adventuresome fieldwork. My point in describing the two trunks and their respective rootstocks is merely to sketch a range of possibilities — and maybe also to provoke some fruitful argument.

Several other viny stalks deserve mentioning. One is straightforward science reporting as practiced by well-informed, fastidious journalists and authors who are not current participants in the scientific enterprise, though they may have had scientific training. Where did that lineage begin? I couldn't say, though possibly William L. Lawrence's wartime reports to the *New York Times* from the Manhattan Project were among the early models. The big issues at Los Alamos and Oak Ridge were technological, not scientific, it's true, but revolutionary insights in nuclear physics lay just below the surface of each day's engineering experiments. And to follow the lanky strides of Robert Oppenheimer while scribbling on a notepad must have been one of the more pungent journalistic — and literary — opportunities a writer could ever have. Nowadays we find solid science writing in the "Science Times" and other feature sections of good newspapers, in the more general sections of certain scientific journals (such as *Science,* which ran the Taubes string theory report), and occasionally in magazines and books. Among the best instances of that craft within recent memory is *The Eighth Day of Creation,* Horace Freeland Judson's magisterial 1979 history of the foundational phase of molecular biology.

Another branch of the vine-tree that has thickened nicely in this country is the science essay, especially as delivered by working scientists who, in stolen hours, ruminate on areas of scientific specialty other than their own. This genre goes back at least to the short, wry, illuminating essays that J.B.S. Haldane wrote for the *Daily Worker* in England in the 1930s and 1940s, and to Loren Eiseley's slightly later contributions to *Harper's* and other magazines. Lewis Thomas took the science essay to a high level of humane grace in the *New England Journal of Medicine.* Stephen Jay Gould combined wit and freshness (especially in his earlier work) with scientific brilliance and an encyclopedic knowledge of popular culture to become America's favorite science pundit through his pieces in *Natural History.* Jared Diamond, Robert S. Desowitz, Alan Lightman, and others continue to show us that a sharp scientist with an amiable voice is worth reading regularly on almost any topic.

This clambering vine-tree includes one other stalk that I want to try to delineate. I'm slightly less able to view it dispassionately, since it encompasses the work of my closest professional colleagues and most of my own, too. For a definition of this group, I have to start with negatives: They're not scientists, they're not reporters, they're not nature writers of the decorous, dependable sort. Their hearts are not always so gentle, and mostly they don't stay home. Some critics have accused them of being a sort of sylvan brigade of the New Journalism, whatever that means. They don't have a collective program, any more than the other clusters of writers I've mentioned do, but they share certain interests and affinities. They care about landscape and its history. They care about the diversity of life that occupies landscape. They are not rooted to academic institutions or news organizations, generally, but will travel almost anywhere a magazine cares to send them. They appreciate (some more manifestly than others) the inherent value and literary usefulness of humor. They're prepared to commit irony. They know that politics, though boring, is inescapable and crucial. They are fascinated by bizarre species of all stripes, but especially by *Homo sapiens*. For better and sometimes for worse, they tend to throw an entire personhood into the literary task, not content to maintain the separation of subject from author, or of science and nature from life. They do research. They do legwork. They interview, they eavesdrop, they tag along on expeditions and invade laboratories. With a sense of construction in the spirit of Robert Rauschenberg, they take liberties of untoward juxtaposition and inclusion.

For a landmark example of such work, think of Edward Abbey's *Desert Solitaire,* published in 1968, when it seemed unfashionable and was virtually ignored, recognized only later as an underground, and eventually an aboveground, classic. It's a book of many moods, many attitudes, many narrative and expository bits, united only by the voice of one man and his conviction that desert landscape is vital. Abbey was not any god to the cohort of writers I'm describing, but he was a friend to some, an influence to others, and he remains the charming, flawed, dead older brother in whose shadow they walk. A sampling of other books in this vein might include Charles Bowden's *Blue Desert,* Terry Tempest Williams's *Refuge,* Edward Hoagland's *The Courage of Turtles,* James Gleick's *Chaos,* Peter Matthiessen's *The Snow Leopard,* Jonathan Raban's *Bad Land,* Annie Dillard's *Pilgrim at Tinker Creek,* Bruce Chatwin's *The Songlines,* Doug Peacock's *Grizzly Years,* John McPhee's *Coming into the Country,* Roger Lewin's *Complexity,* Gary Nabhan's *The Desert Smells Like Rain,* Dennis Overbye's *Lonely Hearts of the Cosmos,* William Kittredge's *Hole in the Sky,* Caroline Alexander's *One Dry Season,* Eugene Linden's *Silent Partners,* Tim Cahill's *Jaguars Ripped My Flesh,* Ian Frazier's *Great Plains,* Jonathan Evan Maslow's *Bird of Life, Bird of Death,* Sy Montgomery's *Spell of the Tiger,* Bill Barich's *Laughing in the Hills,* Tom Miller's *On the Border,* and Barry Lopez's *Of Wolves and Men* and *Arctic Dreams.* There are more, including some nifty works by friends and compeers of mine who are secure enough to be unconcerned that I haven't mentioned them explicitly.

It's not a club. It's not a movement. It's just an infectious itch, a common realization that new effects and purposes might be achieved by ignoring old guidelines and limits.

Sometimes, late in the evening, I talk about all this by phone with Barry Lopez, an important colleague to me and a good pal, who shares my sense that there's a fresh smell on the wind, and who confounds the clumsy categories outlined earlier in that he Goes Forth intrepidly all over the planet with what seems to me the most Gentle of Hearts. (His absence from this volume, by the way, merely reflects that he spent 1999 engrossed in work that didn't include any short pieces appropriate for such a collection. The same sort of circumstance holds for a number of other writers, whom I suspect you may see in future installments of the series.) Labels aren't essential, but they are sometimes convenient for calling attention to trends, and one of the idle questions Barry and I swat back and forth is, If it's not nature writing and it's not science journalism and it's not travel writing or social commentary, then what should one call this stuff? "Landscape nonfiction" is a possibility, though it doesn't satisfy either of us. "Political ornithology" is another, an ironic shorthand coined years ago (by the novelist Graeme Gibson, from whom Barry heard it) in admiring reference to Maslow's *Bird of Life, Bird of Death,* a startling book about birdwatching and genocide, resplendent quetzals and vultures among the forests and the body dumps of war-torn Guatemala. We've never found the perfect phrase, Barry and I, in those late-night chats, but so be it. Convenient or not, categorical labels make people chafe. So we now devote our conversations mostly to the state of the planet's landscape, the extinction of cultural diversity as well as species, the ethics and techniques of nonfiction writing, the thrills and travails of the road, the found jewels among whatever we've lately read, life after age fifty, and the weighty conundrum of whether Pete Rose should be admitted to the Baseball Hall of Fame. Writers should just write, I suppose, one piece at a time, doing the best they can by their lights with each chance and each subject and each bit of material, making shapes that entertain and inform and mean, leaving critics and publishers to worry about truth in labeling.

As science has gotten bigger and more potent, artful nonfiction that examines its workings has grown in abundance and significance too. As nature (in the narrow, green sense) has become ever more marginalized and besieged, nature writing has had to reinvent itself, with a sense of outrage, and of outreach, and of dark, desperate humor. The task of writers who care about one or both of these vast subjects is, among other things, to retain a relentless urge for connectedness and a rogue disregard for boundaries. The task of readers is to demand the world.

DAVID QUAMMEN

NATALIE ANGIER

Men, Women, Sex, and Darwin

FROM *The New York Times Magazine*

LIFE IS SHORT, but jingles are forever. None more so, it seems, than the familiar ditty written by William James: "Hoggamus, higgamus,/Men are polygamous,/Higgamus, hoggamus,/Women monogamous."

Lately the pith of that jingle has found new fodder and new fans through the explosive growth of a field known as evolutionary psychology. Evolutionary psychology professes to have discovered the fundamental modules of human nature, most notably the essential nature of man and of woman. It makes sense to be curious about the evolutionary roots of human behavior. It's reasonable to try to understand our impulses and actions by applying Darwinian logic to the problem. We're animals. We're not above the rude little prods and jests of natural and sexual selection. But evolutionary psychology as it has been disseminated across mainstream consciousness is a cranky and despotic Cyclops, its single eye glaring through an overwhelmingly masculinist lens. I say "masculinist" rather than "male" because the view of male behavior promulgated by hard-core evolutionary psychologists is as narrow and inflexible as their view of womanhood is.

I'm not interested in explaining to men what they really want or how they should behave. If a fellow chooses to tell himself that his yen for the fetching young assistant in his office and his concomitant disgruntlement with his aging wife make perfect Darwinian sense, who am I to argue with him? I'm only proposing here that the hard-core evolutionary psychologists have got a lot about women wrong — about some of us, anyway — and that women want more and deserve better than the cartoon Olive Oyl handed down for popular consumption.

The cardinal premises of evolutionary psychology of interest to this discussion are as follows: 1. Men are more promiscuous and less sexually reserved than women are. 2. Women are inherently more interested in a stable relationship than men are. 3. Women are naturally attracted to high-status men with resources. 4. Men are naturally attracted to youth and beauty. 5. Humankind's core preferences and desires were hammered out long, long ago, a hundred thousand years or more, in the legendary Environment of Evolutionary Adaptation, or EEA, also known as the ancestral environment, also known as the Stone Age, and they have not changed appreciably since then, nor are they likely to change in the future.

In sum: Higgamus, hoggamus, Pygmalionus, *Playboy* magazine, eternitas. Amen.

Hard-core evolutionary psychology types go to extremes to argue in favor of the yawning chasm that separates the innate desires of women and men. They declare ringing confirmation for their theories even in the face of feeble and amusingly contradictory data. For example: Among the cardinal principles of the evo-psycho set is that men are by nature more polygamous than women are, and much more accepting of casual, even anonymous, sex. Men can't help themselves, they say: they are always hungry for sex, bodies, novelty, and nubility. Granted, men needn't act on such desires, but the drive to sow seed is there nonetheless, satyric and relentless, and women cannot fully understand its force. David Buss, a professor of psychology at the University of Texas at Austin and one of the most outspoken of the evolutionary psychologists, says that asking a man not to lust after a pretty young woman is like telling a carnivore not to like meat.

At the same time, they recognize that the overwhelming majority of men and women get married, and so their theories must extend to different innate mate preferences among men and women. Men look for the hallmarks of youth, like smooth skin, full lips, and perky breasts; they want a mate who has a long childbearing career ahead of her. Men also want women who are virginal and who seem as though they'll be faithful and not make cuckolds of them. The sexy, vampy types are fine for a Saturday romp, but when it comes to choosing a marital partner, men want modesty and fidelity.

Women want a provider, the theory goes. They want a man who

seems rich, stable, and ambitious. They want to know that they and their children will be cared for. They want a man who can take charge, maybe dominate them just a little, enough to reassure them that the man is genotypically, phenotypically, eternally, a king. Women's innate preference for a well-to-do man continues to this day, the evolutionary psychologists insist, even among financially independent and professionally successful women who don't need a man as a provider. It was adaptive in the past to look for the most resourceful man, they say, and adaptations can't be willed away in a generation or two of putative cultural change.

And what is the evidence for these male-female verities? For the difference in promiscuity quotas, the hard-cores love to raise the example of the differences between gay men and lesbians. Homosexuals are seen as a revealing population because they supposedly can behave according to the innermost impulses of their sex, untempered by the need to adjust to the demands and wishes of the opposite sex, as heterosexuals theoretically are. What do we see in this ideal study group? Just look at how gay men carry on! They are perfectly happy to have hundreds, thousands, of sexual partners, to have sex in bathhouses, in bathrooms, in Central Park. By contrast, lesbians are sexually sedate. They don't cruise sex clubs. They couple up and stay coupled, and they like cuddling and hugging more than they do serious, genitally based sex.

In the hard-core rendering of inherent male-female discrepancies in promiscuity, gay men are offered up as true men, real men, men set free to be men, while lesbians are real women, ultrawomen, acting out every woman's fantasy of love and commitment. Interestingly, though, in many neurobiology studies gay men are said to have somewhat feminized brains, with hypothalamic nuclei that are closer in size to a woman's than to a straight man's, and spatial-reasoning skills that are modest and ladylike rather than manfully robust. For their part, lesbians are posited to have somewhat masculinized brains and skills — to be sportier, more mechanically inclined, less likely to have played with dolls or tea sets when young — all as an ostensible result of exposure to prenatal androgens. And so gay men are sissy boys in some contexts and Stone Age manly men in others, while lesbians are battering rams one day and flower into the softest and most sexually divested girlish girls the next.

On the question of mate preferences, evo-psychos rely on sur-

veys, most of them compiled by David Buss. His surveys are celebrated by some, derided by others, but in any event they are ambitious — performed in thirty-seven countries, he says, on six continents. His surveys, and others emulating them, consistently find
that men rate youth and beauty as important traits in a mate,
while women give comparatively greater weight to ambition and
financial success. Surveys show that surveys never lie. Lest you
think that women's mate preferences change with their own
mounting economic clout, surveys assure us that they do not. Surveys of female medical students, according to John Marshall
Townsend, of Syracuse University, indicate that they hope to marry
men with an earning power and social status at least equal to and
preferably greater than their own.

Perhaps all this means is that men can earn a living wage better,
even now, than women can. Men make up about half the world's
population, but they still own the vast majority of the world's wealth
— the currency, the minerals, the timber, the gold, the stocks, the
amber fields of grain. In her superb book *Why So Slow?* Virginia
Valian, a professor of psychology at Hunter College, lays out the extent of lingering economic discrepancies between men and women
in the United States. In 1978 there were two women heading Fortune 1000 companies; in 1994, there were still two; in 1996, the
number had jumped all the way to four. In 1985, 2 percent of the
Fortune 1000's senior-level executives were women; by 1992, that
number had hardly budged, to 3 percent. A 1990 salary and compensation survey of 799 major companies showed that of the high-
est-paid officers and directors, less than one half of 1 percent were
women. Ask, and he shall receive. In the United States the possession of a bachelor's degree adds $28,000 to a man's salary but
only $9,000 to a woman's. A degree from a high-prestige school
contributes $11,500 to a man's income but subtracts $2,400 from
a woman's. If women continue to worry that they need a man's
money, because the playing field remains about as level as the surface of Mars, then we can't conclude anything about innate preferences. If women continue to suffer from bag-lady syndrome even as
they become prosperous, if they still see their wealth as provisional
and capsizable, and if they still hope to find a man with a dependable income to supplement their own, then we can credit women
with intelligence and acumen, for inequities abound.

There's another reason that smart, professional women might respond on surveys that they'd like a mate of their socioeconomic status or better. Smart, professional women are smart enough to know that men can be tender of ego — is it genetic? — and that it hurts a man to earn less money than his wife, and that resentment is a noxious chemical in a marriage and best avoided at any price. "A woman who is more successful than her mate threatens his position in the male hierarchy," Elizabeth Cashdan, of the University of Utah, has written. If women could be persuaded that men didn't mind their being high achievers, were in fact pleased and proud to be affiliated with them, we might predict that the women would stop caring about the particulars of their mates' income. The anthropologist Sarah Blaffer Hrdy writes that "when female status and access to resources do not depend on her mate's status, women will likely use a range of criteria, not primarily or even necessarily prestige and wealth, for mate selection." She cites a 1996 *New York Times* story about women from a wide range of professions — bankers, judges, teachers, journalists — who marry male convicts. The allure of such men is not their income, for you can't earn much when you make license plates for a living. Instead, it is the men's gratitude that proves irresistible. The women also like the fact that their husbands' fidelity is guaranteed. "Peculiar as it is," Hrdy writes, "this vignette of sex-reversed claustration makes a serious point about just how little we know about female choice in breeding systems where male interests are not paramount and patrilines are not making the rules."

Do women love older men? Do women find gray hair and wrinkles attractive on men — as attractive, that is, as a fine, full head of pigmented hair and a vigorous, firm complexion? The evolutionary psychologists suggest yes. They believe that women look for the signs of maturity in men because a mature man is likely to be a comparatively wealthy and resourceful man. That should logically include baldness, which generally comes with age and the higher status that it often confers. Yet, as Desmond Morris points out, a thinning hairline is not considered a particularly attractive state.

Assuming that women find older men attractive, is it the men's alpha status? Or could it be something less complimentary to the male, something like the following: that an older man is appealing not because he is powerful but because in his maturity he has lost some of his power, has become less marketable and desirable and potentially more grateful and gracious, more likely to make a younger woman feel that there is a balance of power in the relationship? The rude little calculation is simple: He is male, I am female — advantage, man. He is older, I am younger — advantage, woman. By the same token, a woman may place little value on a man's appearance because she values something else far more: room to breathe. Who can breathe in the presence of a handsome young man, whose ego, if expressed as a vapor, would fill Biosphere II? Not even, I'm afraid, a beautiful young woman.

In the end, what is important to question, and to hold to the fire of alternative interpretation, is the immutability and adaptive logic of the discrepancy, its basis in our genome rather than in the ecological circumstances in which a genome manages to express itself. Evolutionary psychologists insist on the essential discordance between the strength of the sex drive in males and females. They admit that many nonhuman female primates gallivant about rather more than we might have predicted before primatologists began observing their behavior in the field — more, far more, than is necessary for the sake of reproduction. Nonetheless, the credo of the coy female persists. It is garlanded with qualifications and is admitted to be an imperfect portrayal of female mating strategies, but then, that little matter of etiquette attended to, the credo is stated once again.

"Amid the great variety of social structure in these species, the basic theme . . . stands out, at least in minimal form: males seem very eager for sex and work hard to find it; females work less hard," Robert Wright says in *The Moral Animal.* "This isn't to say the females don't like sex. They love it, and may initiate it. And, intriguingly, the females of the species most closely related to humans — chimpanzees and bonobos — seem particularly amenable to a wild sex life, including a variety of partners. Still, female apes don't do what male apes do: search high and low, risking life and limb, to find sex, and to find as much of it, with as many different partners, as possible; it has a way of finding them." In fact, female chimpanzees do seem to search high and low and take great risks to find sex with partners other than the partners who have a way of finding them. DNA studies of chimpanzees in West Africa show that half the offspring in a group of closely scrutinized chimpanzees turned out not to be the offspring of the resident males. The females of the group didn't rely on sex "finding" its way to them; they proactively left the local environs, under such conditions of secrecy that not even their vigilant human observers knew they had gone, and became impregnated by outside males. They did so even at the risk of life and limb — their own and those of their offspring. Male chimpanzees try to control the movements of fertile females. They'll scream at them and hit them if they think the females aren't listening. They may even kill an infant they think is not their own. We don't know why the females take such risks to philander, but they do, and to say that female chimpanzees "work less hard" than males do at finding sex does not appear to be supported by the data.

Evo-psychos pull us back and forth until we might want to sue for whiplash. On the one hand, we are told that women have a lower sex drive than men do. On the other hand, we are told that the madonna-whore dichotomy is a universal stereotype. In every culture there is a tendency among both men and women to adjudge women as either chaste or trampy. The chaste ones are accorded esteem. The trampy ones are consigned to the basement, a notch or two below goats in social status. A woman can't sleep around without risking terrible retribution — to her reputation, to her prospects, to her life. "Can anyone find a single culture in which women with unrestrained sexual appetites aren't viewed as more aberrant than comparably libidinous men?" Wright asks rhetorically.

Women are said to have lower sex drives than men, yet they are universally punished if they display evidence to the contrary — if they disobey their "natural" inclination toward a stifled libido. Women supposedly have a lower sex drive than men do, yet it is not low enough. There is still just enough of a lingering female infidelity impulse that cultures everywhere have had to gird against it by articulating a rigid dichotomy with menacing implications for those who fall on the wrong side of it. There is still enough lingering female infidelity to justify infibulation, purdah, claustration. Men have the naturally higher sex drive, yet all the laws, customs, punishments, shame, strictures, mystiques, and antimystiques are aimed with full hominid fury at that tepid, sleepy, hypoactive creature, the female libido.

"It seems premature . . . to attribute the relative lack of female interest in sexual variety to women's biological nature alone in the face of overwhelming evidence that women are consistently beaten for promiscuity and adultery," the primatologist Barbara Smuts has written. "If female sexuality is muted compared to that of men, then why must men the world over go to extreme lengths to control and contain it?"

Why indeed? Consider a brief evolutionary apologia for President Clinton's adulteries written by Steven Pinker, of the Massachusetts Institute of Technology. "Most human drives have ancient Darwinian rationales," he wrote. "A prehistoric man who slept with fifty women could have sired fifty children, and would have been more likely to have descendants who inherited his tastes. A woman who slept with fifty men would have no more descendants than a woman who slept with one. Thus, men should seek quantity in sexual partners; women, quality." And isn't it so, he says, everywhere and always so? "In our society," he continues, "most young men tell researchers that they would like eight sexual partners in the next two years; most women say that they would like one." Yet would a man find the prospect of a string of partners so appealing if the following rules were applied: that no matter how much he may like a particular woman and be pleased by her performance and want to sleep with her again, he will have no say in the matter and will be dependent on her mood and good graces for all future contact; that each act of casual sex will cheapen his status and make him increasingly less attractive to other women; and that society will not wink at his randiness but rather sneer at him and think him pathetic, sullied, smaller than life? Until men are subjected to the same severe standards and threat of censure as women are, and until they are given the lower hand in a so-called casual encounter from the start, it is hard to insist with such self-satisfaction that, hey, it's natural, men like a lot of sex with a lot of people and women don't.

Reflect for a moment on Pinker's philandering caveman who slept with fifty women. Just how good a reproductive strategy is this chronic, random shooting of the gun? A woman is fertile only five or six days a month. Her ovulation is concealed. The man doesn't know when she's fertile. She might be in the early stages of pregnancy when he gets to her; she might still be lactating and thus not ovulating. Moreover, even if our hypothetical Don Juan hits a day on which a woman is ovulating, the chances are around 65 percent that his sperm will fail to fertilize her egg; human reproduction is complicated, and most eggs and sperm are not up to the demands of proper fusion. Even if conception occurs, the resulting embryo has about a 30 percent chance of miscarrying at some point in gestation. In sum, each episode of fleeting sex has a remarkably small probability of yielding a baby — no more than 1 or 2 percent, at best.

And because the man is trysting and running, he isn't able to prevent any of his casual contacts from turning around and mating with other men. The poor fellow. He has to mate with many scores of women for his wham-bam strategy to pay off. And where are all these women to be found, anyway? Population densities during that purportedly all-powerful psyche shaper the "ancestral environment" were quite low, and long-distance travel was dangerous and difficult.

There are alternatives to wantonness, as a number of theorists have emphasized. If, for example, a man were to spend more time with one woman rather than dashing breathlessly from sheet to sheet, if he were to feel compelled to engage in what animal behaviorists call mate guarding, he might be better off, reproductively speaking, than the wild Lothario, both because the odds of impregnating the woman would increase and because he'd be monopolizing her energy and keeping her from the advances of other sperm bearers. It takes the average couple three to four months of regular sexual intercourse to become pregnant. That number of days is approximately equal to the number of partners our hypothetical libertine needs to sleep with to have one encounter result in a "fertility unit," that is, a baby. The two strategies, then, shake out about the same. A man can sleep with a lot of women — the quantitative approach — or he can sleep with one woman for months at a time, and be madly in love with her — the qualitative tactic.

It's possible that these two reproductive strategies are distributed in discrete packets among the male population, with a result that some men are born philanderers and can never attach, while others are born romantics and perpetually in love with love. But it's also possible that men teeter back and forth from one impulse to the other, suffering an internal struggle between the desire to bond and the desire to retreat, with the circuits of attachment ever there to be toyed with, and their needs and desires difficult to understand, paradoxical, fickle, treacherous, and glorious. It is possible, then, and for perfectly good Darwinian reasons, that casual sex for men is rarely as casual as it is billed.

It needn't be argued that men and women are exactly the same, or that humans are meta-evolutionary beings, removed from nature and slaves to culture, to reject the perpetually regurgitated model of the coy female and the ardent male. Conflicts of interest are always among us, and the outcomes of those conflicts are interesting, more interesting by far than what the ultra-evolutionary psychology line has handed us. Patricia Gowaty, of the University of Georgia, sees conflict between males and females as inevitable and pervasive. She calls it sexual dialectics. Her thesis is that females and males vie for control over the means of reproduction. Those means are the female body, for there is as yet no such beast as the parthenogenetic man.

Women are under selective pressure to maintain control over their reproduction, to choose with whom they will mate and with whom they will not — to exercise female choice. Men are under selective pressure to make sure they're chosen or, barring that, to subvert female choice and coerce the female to mate against her will. "But once you have this basic dialectic set in motion, it's going to be a constant push-me, pull-you," Gowaty says. That dynamism cannot possibly result in a unitary response, the caricatured coy woman and ardent man. Instead there are going to be some coy, reluctantly mating males and some ardent females, and any number of variations in between.

"A female will choose to mate with a male whom she believes, consciously or otherwise, will confer some advantage on her and her offspring. If that's the case, then her decision is contingent on what she brings to the equation." For example, she says, "the 'good genes' model leads to oversimplified notions that there is a 'best male' out there, a top-of-the-line hunk whom all females would prefer to mate with if they had the wherewithal. But in the viability model, a female brings her own genetic complement to the equation, with the result that what looks good genetically to one woman might be a clash of colors for another."

Maybe the man's immune system doesn't complement her own,

for example, Gowaty proposes. There's evidence that the search for immune variation is one of the subtle factors driving mate selection, which may be why we care about how our lovers smell; immune molecules may be volatilized and released in sweat, hair, the oil on our skin. We are each of us a chemistry set, and each of us has a distinctive mix of reagents. "What pleases me might not please somebody else," Gowaty says. "There is no one-brand great male out there. We're not all programmed to look for the alpha male and only willing to mate with the little guy or the less aggressive guy because we can't do any better. But the propaganda gives us a picture of the right man and the ideal woman, and the effect of the propaganda is insidious. It becomes self-reinforcing. People who don't fit the model think, I'm weird, I'll have to change my behavior." It is this danger, that the ostensible "discoveries" of evolutionary psychology will be used as propaganda, that makes the enterprise so disturbing.

Variation and flexibility are the key themes that get set aside in the breathless dissemination of evolutionary psychology. "The variation is tremendous, and is rooted in biology," Barbara Smuts said to me. "Flexibility itself is the adaptation." Smuts has studied olive baboons, and she has seen males pursuing all sorts of mating strategies. "There are some whose primary strategy is dominating other males, and being able to gain access to more females because of their fighting ability," she says. "Then there is the type of male who avoids competition and cultivates long-term relationships with females and their infants. These are the nice, affiliative guys. There's a third type, who focuses on sexual relationships. He's the consorter. . . . And as far as we can tell, no one reproductive strategy has advantages over the others."

Women are said to need an investing male. We think we know the reason. Human babies are difficult and time-consuming to raise. Stone Age mothers needed husbands to bring home the bison. Yet the age-old assumption that male parental investment lies at the heart of human evolution is now open to serious question. Men in traditional foraging cultures do not necessarily invest resources in their offspring. Among the Hadza of Africa, for example, the men hunt, but they share the bounty of that hunting widely, politically, strategically. They don't deliver it straight to the mouths of their progeny. Women rely on their senior female kin to help feed their children. The women and their children in a gathering-hunting society clearly benefit from the meat that hunters bring back to the group. But they benefit as a group, not as a collection of nuclear family units, each beholden to the father's personal pound of wildeburger.

This is a startling revelation, which upends many of our presumptions about the origins of marriage and what women want from men and men from women. If the environment of evolutionary adaptation is not defined primarily by male parental investment, the bedrock of so much of evolutionary psychology's theories, then we can throw the door wide open and ask new questions, rather than endlessly repeating ditties and calling the female coy long after she has run her petticoats through the presidential paper shredder.

For example: Nicholas Blurton Jones, of the University of California at Los Angeles, and others have proposed that marriage developed as an extension of men's efforts at mate guarding. If the cost of philandering becomes ludicrously high, the man might be better off trying to claim rights to one woman at a time. Regular sex with a fertile woman is at least likely to yield offspring at comparatively little risk to his life, particularly if sexual access to the woman is formalized through a public ceremony — a wedding. Looked at from this perspective, one must wonder why an ancestral woman bothered to get married, particularly if she and her female relatives did most of the work of keeping the family fed from year to year. Perhaps, Blurton Jones suggests, to limit the degree to which she was harassed. The cost of chronic male harassment may be too high to bear. Better to agree to a ritualized bond with a male, and to benefit from whatever hands-off policy that marriage may bring, than to spend all of her time locked in one sexual dialectic or another.

Thus marriage may have arisen as a multifaceted social pact: between man and woman, between male and male, and between the couple and the tribe. It is a reasonable solution to a series of cultural challenges that arose in concert with the expansion of the human neocortex. But its roots may not be what we think they are, nor may our contemporary mating behaviors stem from the pressures of an ancestral environment as it is commonly portrayed, in which a woman needed a mate to help feed and clothe her young.

Instead, our "deep" feelings about marriage may be more pragmatic, more contextual, and, dare I say it, more egalitarian than we give them credit for being.

If marriage is a social compact, a mutual bid between man and woman to contrive a reasonably stable and agreeable microhabitat in a community of shrewd and well-armed members, then we can understand why, despite rhetoric to the contrary, men are as eager to marry as women are. A raft of epidemiological studies have shown that marriage adds more years to the life of a man than it does to that of a woman. Why should that be, if men are so "naturally" ill suited to matrimony?

What do women want? None of us can speak for all women, or for more than one woman, really, but we can hazard a mad guess that a desire for emotional parity is widespread and profound. It doesn't go away, although it often hibernates under duress, and it may be perverted by the restrictions of habitat or culture into something that looks like its opposite. The impulse for liberty is congenital. It is the ultimate manifestation of selfishness, which is why we can count on its endurance.

WENDELL BERRY

Back to the Land

FROM *The Amicus Journal*

ONE OF THE PRIMARY results — and one of the primary needs
— of industrialism is the separation of people and places and products from their histories. To the extent that we participate in the industrial economy, we do not know the histories of our families or of
our habitats or of our meals. This is an economy, and in fact a culture, of the one-night stand. "I had a good time," says the industrial
lover, "but don't ask me my last name." Just so, the industrial eater
says to the svelte industrial hog, "We'll be together at breakfast. I
don't want to see you before then, and I won't care to remember
you afterward."

In this condition, we have many commodities but little satisfaction, little sense of the sufficiency of anything. The scarcity of
satisfaction makes of our many commodities an infinite series of
commodities, the new commodities invariably promising greater
satisfaction than the older ones. In fact, the industrial economy's
most marketed commodity is satisfaction, and this commodity,
which is repeatedly promised, bought, and paid for, is never delivered.

This persistent want of satisfaction is directly and complexly related to the dissociation of ourselves and all our goods from our
and their histories. If things do not last, are not made to last, they
can have no histories, and we who use these things can have no
memories. One of the procedures of the industrial economy is to
reduce the longevity of materials. For example, wood — which,
well made into buildings and furniture and well cared for, can last
hundreds of years — is now routinely manufactured into products

that last just twenty-five years. We do not cherish the memory of shoddy and transitory objects, and so we do not remember them. That is to say that we do not invest in them the lasting respect and admiration that make for satisfaction.

The problem of our dissatisfaction with all the things that we use is not correctable within the terms of the economy that produces those things. At present, it is virtually impossible for us to know the economic history or the ecological cost of the products we buy; the origins of the products are typically too distant and too scattered and the processes of trade, manufacture, transportation, and marketing too complicated. There are, moreover, too many good reasons for the industrial suppliers of these products not to want their histories to be known.

Where there is no reliable accounting and therefore no competent knowledge of the economic and ecological effects of our lives, we cannot live lives that are economically and ecologically responsible. This is the problem that has frustrated, and to a considerable extent undermined, the American conservation effort from the beginning. It is ultimately futile to plead and protest and lobby in favor of public ecological responsibility while, in virtually every act of our private lives, we endorse and support an economic system that is by intention, and perhaps by necessity, ecologically irresponsible.

If the industrial economy is not correctable within or by its own terms, then obviously what is required for correction is a countervailing economic idea. And the most significant weakness of the conservation movement is its failure to produce or espouse an economic idea capable of correcting the economic idea of the industrialists. Anybody who has studied with care the issues of conservation knows that our acts are being measured by a real and unyielding standard that was invented by no human. Our acts that are not in harmony with nature are inevitably and sometimes irremediably destructive. The standard exists. But having no opposing economic idea, the conservationists have had great difficulty in applying the standard.

What, then, is the countervailing idea by which we might correct the industrial idea? We will not have to look hard to find it, for there is only one, and that is agrarianism. Our major difficulty (and danger) will be in attempting to deal with agrarianism as "an idea"

— agrarianism is primarily a practice, a set of attitudes, a loyalty, and a passion; it is an idea only secondarily and at a remove. I am well aware of the danger in defining things, but if I am going to talk about agrarianism, I am going to have to define it. The definition that follows is derived both from agrarian writers, ancient and modern, and from the unliterary and sometimes illiterate agrarians who have been my teachers.

The fundamental difference between industrialism and agrarianism is this: Whereas industrialism is a way of thought based on monetary capital and technology, agrarianism is a way of thought based on land.

An agrarian economy rises up from the fields, woods, and streams — from the complex of soils, slopes, weathers, connections, influences, and exchanges that we mean when we speak, for example, of the local community or the local watershed. The agrarian mind is therefore not regional or national, let alone global, but local. It must know on intimate terms the local plants and animals and local soils; it must know local possibilities and impossibilities, opportunities and hazards.

Because a mind so placed meets again and again the necessity for work to be good, the agrarian mind is less interested in abstract quantities than in particular qualities. It feels threatened and sickened when it hears people and creatures and places spoken of as labor, management, capital, and raw material. It is not at all impressed by the industrial legendry of gross national products or of the numbers sold and dollars earned by gigantic corporations. It is interested in, and forever fascinated by, questions leading toward the accomplishment of good work: What is the best location for a particular building or fence? What is the best way to plow this field? What is the best course for a skid road in this woodland? Should this tree be cut or spared? — questions that cannot be answered in the abstract, and that yearn not toward quantity but toward elegance. Agrarianism can never become abstract, because it has to be practiced in order to exist.

An agrarian economy is always a subsistence economy before it is a market economy. The center of an agrarian farm is the household. It is the subsistence part of the agrarian economy that assures its stability and its survival. A subsistence economy necessarily is highly diversified, and it characteristically has involved hunting and gathering as well as farming and gardening. These activities bind people to their local landscape by close, complex interests and economic ties.

The stability, coherence, and longevity of human occupation require that the land should be divided among many owners and users. The central figure of agrarian thought has invariably been the small owner or small holder who maintains a significant measure of economic self-determination on a small acreage. The scale and independence of such holdings imply two things that agrarians see as desirable: intimate care in the use of the land and political democracy resting upon the indispensable foundation of economic democracy. A major characteristic of the agrarian mind is a longing for independence — that is, for an appropriate degree of personal and local self-sufficiency. Agrarians wish to earn and deserve what they have. They do not wish to live by piracy, beggary, charity, or luck.

The agrarian mind begins with the love of fields and ramifies in good farming, good cooking, good eating, and gratitude to God. Exactly analogous to the agrarian mind is the sylvan mind, which begins with the love of forests and ramifies in good forestry, good woodworking, good carpentry, etc., and in gratitude to God. These two kinds of mind readily intersect and communicate; neither ever intersects or communicates with the industrial-economic mind. The industrial-economic mind begins with ingratitude and ramifies in the destruction of farms and forests. The "lowly" and "menial" arts of farm and forest are mostly taken for granted or ignored by the culture of the "fine arts" and by "spiritual" religions; they are taken for granted or ignored or held in contempt by the powers of the industrial economy. But in fact they are inescapably the foundation of human life and culture, and their adepts are capable of as deep satisfaction and as high attainments as anybody else.

Having, so to speak, laid industrialism and agrarianism side by side, implying a preference for the latter, I will be confronted by two questions that I had better go ahead and answer.

The first is whether or not agrarianism is simply a "phase" that we humans had to go through and then leave behind in order to get onto the track of technological progress toward ever greater happiness. The answer is that although industrialism has certainly conquered agrarianism, and has very nearly destroyed it altogether, in every one of its uses of the natural world industrialism is in the process of catastrophic failure. Industry is now desperately shifting — by means of genetic engineering, global colonialism, and other contrivances — to prolong its control of our farms and forests, but the failure nonetheless continues. It is not possible to argue sanely in favor of soil erosion, water pollution, genetic impoverishment, and the destruction of rural communities and local economies. Industrialism, unchecked by the affections and concerns of agrarianism, becomes monstrous.

The second question is whether or not by espousing the revival of agrarianism we will commit the famous sin of "turning back the clock." The answer to that, for present-day North Americans, is fairly simple. Agrarian people wish to fit the farming to the farm and the forestry to the forest. At times and in places we latter-day Americans may have come close to accomplishing this goal, but we never yet have developed stable, sustainable, locally adapted land-based economies. The good rural enterprises and communities that we will find in our past have been almost constantly under threat from the colonialism, first foreign and then domestic, that has been institutionalized for a long time in the industrial economy. The possibility of an authentically settled country still lies ahead of us.

If we wish to look ahead, we will see, not only in the United States but in the world, two economic programs that conform pretty exactly to the aims of industrialism and agrarianism as I have described them.

The first is the effort to globalize the industrial economy, not merely by the expansionist programs of supra-national corporations within themselves, but also by means of government-sponsored international trade agreements, the most prominent of which is the General Agreement on Tariffs and Trade.

The second program, counter to the first, comprises many small efforts to preserve or improve or establish local economies. These efforts on the part of nonindustrial or agrarian conservatives, local patriots, are taking place in countries both affluent and poor all over the world.

The global economists are the great centralizers of our time. The local economists, who have so far attracted the support of no prominent politician, are the true decentralizers and downsizers, for they seek an appropriate degree of self-determination and independence for localities. They seem to be moving toward a radical and necessary revision of our idea of a city. They are learning to see the city, not just as a built and paved municipality set apart by "city limits" to live by trade and transportation from the world at large, but rather as a part of a community that includes also the city's rural neighbors, its surrounding landscape, and its watershed, on which it might depend for at least some of its necessities, and for the health of which it might exercise a competent concern and responsibility. For though agrarianism proposes that everybody has agrarian responsibilities, it does not propose that everybody should be a farmer or that we do not need cities. Furthermore, any thinkable human economy would have to grant to manufacturing an appropriate and honorable place. Agrarians would insist only that any manufacturing enterprise should be formed and scaled to fit the local landscape, the local ecosystem, and the local community, and that it should be locally owned and employ local people.

Between these two programs — the industrial and the agrarian, the global and the local — the most critical difference is that of knowledge. The global economy institutionalizes a global ignorance, in which producers and consumers cannot know or care about one another, and in which the histories of all products will be lost.

But in a sound local economy, in which producers and consumers are neighbors, nature will become the known standard of work and production. Consumers who understand their economy will not tolerate the destruction of the local soil or ecosystem or watershed as a cost of production. Only a healthy local economy can keep nature and work together in the consciousness of the community. Only such a community can restore history to economics.

What agrarian principles implicitly propose — and what I explicitly propose in advocating these principles at this time — is a revolt of local small producers and local consumers against the global industrialism of the corporations. Do I think that there is a hope that such a revolt can survive and succeed, and that it can have a significant influence upon our lives and our world?

Yes, I do. And to be as clear as possible in arguing for this hope, let me begin with an example. Not long ago I received a phone call from my friend David Kline in Holmes County, Ohio. David is an Amish minister, one of the best farmers I know, and a man of excellent judgment. He told me, among other things, that he and his neighbors are now selling organic milk at a premium of several cents a pound above the price of nonorganic milk, and that a small cheese factory in their community is about to begin marketing organic cheese. He said that industrial excesses and abuses in milk production are "making the market" for these organic products. As a result, the farm economy has improved in Holmes County — where, because of the Amish economic and agricultural practices, the farm economy has been pretty good anyhow.

This, I think, gives the pattern of an economic revolt that not only is possible but is in fact happening. It is happening for two reasons: First, as the scale of industrial agriculture increases, so does the scale of its abuses, and it is hard to hide large-scale abuses from consumers. Second, as the food industries focus more and more on gigantic global opportunities, they cannot help but overlook small local opportunities, as is made plain by the increase of "community-supported agriculture," farmers' markets, health-food stores, and so on. In fact, there are some markets that the great corporations by definition cannot supply. The market for so-called organic food, for example, is really a market for good, fresh, trustworthy food, food from producers known and trusted by consumers, and such food cannot be produced by a global corporation.

But the food economy is only one example. It is also possible to think of good local forest economies. And in the face of much neglect, it is possible to think of local small-business economies — some of them related to the local economies of farm and forest — supported by locally owned, community-oriented banks.

What do these struggling, sometimes failing, sometimes hardly realized efforts of local economy have to do with conservation as we know it? The answer, probably, is everything.

I would like my fellow conservationists to notice how many people and organizations are now working to save something of value — not just wilderness places, wild rivers, wildlife habitats, species diversity, water quality, and air quality, but also agricultural land, family farms and ranches, communities, children and childhood, local schools, local economies, local food markets, livestock breeds and domestic plant varieties, fine old buildings, scenic roads, and so on. I would like my fellow conservationists to understand also that there is hardly a small farm or ranch or locally owned restaurant or store or shop or business anywhere that is not struggling to save itself.

All of these people, who are fighting sometimes lonely battles to preserve things of value that they cannot bear to lose, are the conservation movement's natural allies. Most of them have the same enemies as the conservation movement. There is no necessary conflict among them. Thinking of them, in their great variety, in the essential likeness of their motives and concerns, one thinks almost automatically of the possibility of a defined community of interest among them all, a shared stewardship of all the diversity of good things that are needed for the health and abundance of the world.

I don't suppose that this will be easy, given especially the history of conflict between conservationists and land users. I only suppose that it is necessary. Conservationists can't conserve everything that needs conserving without joining the effort to use well the agricultural lands, the forests, and the waters that we must use. To enlarge the areas protected from use without at the same time enlarging the areas of good use is a mistake.

We know better than to expect very soon a working model of a conserving global corporation. But we must begin to expect — and we must, as conservationists, begin working for, and in — working models of conserving local economies. These are possible now. Good and able people are working hard to develop them now. They need the full support of the conservationist movement now. Conservationists need to go to these people, ask what they can do to help, and then help. A little later, having helped, they can in turn ask for help.

RICHARD CONNIFF

Africa's Wild Dogs

FROM *National Geographic*

SOMEWHERE DEEP in Botswana's Okavango Delta, a million miles from nowhere, a dog named Nomad leads his pack on a wild chase through the bush. The sun paints a gaudy orange stripe across the horizon. Night threatens at any moment to rush down and set the lions afoot. Our Land Rover bucks and jumps through a dense thicket of mopani trees, struggling to keep up, then breaks out onto a floodplain through the cat-piss smell of windshield-high sage. Giraffes and tsessebes scatter ahead of us, kicking up panicky clouds of dust. Nomad is the orphan child of a male named Chance and a bitch named Fate, and maybe more sensible men would take the hint and give up, go home, get dinner.

The driver, a wildlife biologist named John "Tico" McNutt, spots a herd of impalas, fast food for the wild dogs we are following. But there are no dogs in sight. He listens to his earphones for the signal from Nomad's radio collar, and then the Land Rover dives back into the bush. "Uh-oh," McNutt says as he muscles the wheel one way and then the other. "Uh-oh." He circles a tree once to get his bearings, then lurches off in the direction that makes his earphones ping strong as a heart monitor. Thorny acacia branches howl down the sides of the truck and leap in at the open windows. A rotten log explodes under our tires, showering us with debris. "Captain, we've been hit!" McNutt reports and guns the engine.

And then we see the dogs out ahead of us, long-legged and light-footed, barely skimming the ground. They stand more than two feet tall at the shoulder, mottled all over with patches of brown, yellow, black, and white. Their ears are round as satellite dishes, and their mouths are slightly open. Everything about them as they glide through the mopanis seems effortless. Then they vanish, dappled shadows moving among the dappled shadows of dusk.

They're commonly called African wild dogs, an unfortunate name suggesting house pets gone bad. In fact, *Lycaon pictus*, the lone species in its genus, is utterly wild and only distantly related to our domestic dog or any other canid. Wild dogs most closely resemble wolves in their social behavior, though they seem more gentle. They are like wolves, too, in that humans have vilified and persecuted them into extinction over most of their range.

Wild dogs once roamed throughout sub-Saharan Africa in every habitat except jungle or desert. A traveler in the 1960s sighted them even in the snows of Mount Kilimanjaro. But they hang on now in just a few isolated pockets, with a total population estimated at fewer than 5,000 animals. They are nearly as endangered as the black rhino, but less celebrated. Farmers still trap them because wild dogs sometimes eat their calves. Hunters occasionally shoot them because they think the dogs steal their game or because they abhor the dogs' reputedly barbaric killing methods. Until the late 1970s even national park managers routinely killed them. The lore was that wild dogs are an "abomination," capable of killing humans, practicing cannibalism, and whenever possible subjecting their prey to a lingering, brutal death. A pack will chase an animal relentlessly, according to various lurid accounts, "tearing away ribbons of skin or lumps of flesh" until the terrified victim "sinks exhausted, when the pack continue to rend out pieces from the living animal."

One misguided hunter dreamed, in 1914, about the "excellent day . . . when means can be devised . . . for this unnecessary creature's complete extermination." Only now, with that day upon us, has it dawned on people that maybe wild dogs aren't so bad after all. They do indeed kill by disemboweling their prey. But death is typically quick and no more barbaric than the noble lion's using its jaws to strangle a flailing zebra. Wild dogs also run in packs, as alleged. But within the pack they practice family values to a degree that would please, or possibly shame, our leading politicians.

The Okavango Delta is a 6,200-square-mile expanse of floodplains and sand ridges, one of the last places in Africa big enough to accommodate wild dogs in their accustomed freedom. Tico McNutt began studying the dogs here in 1989. He is forty-two years old, tall and lean, with blue eyes and a second-day stubble. For nearly ten years McNutt has followed the lives of dozens of wild dog packs and hundreds of individual dogs. He knows almost every dog in his study area by the distinctive mottling of its fur. Often he knows its parents, grandparents, and even great-grandparents as well, allowing him to construct detailed genealogies and observe the rise and fall of dynasties. He names his packs according to theme, and the names sometimes betray longing for his Seattle roots. There are packs named for weather (Typhoon, Tempest, Squall), movie stars (Dustin, Streep, Uma), and beers (Zambezi, Full Sail, Tusker).

He and his wife, Lesley Boggs, an anthropologist specializing in human-wildlife conflict, have written a book, *Running Wild: Dispelling the Myths of the African Wild Dog.* They live in a stand of trees next to a dry floodplain on the edge of the Moremi Game Reserve, in the heart of the Okavango. Their camp is improbably settled and homey, with a basketball hoop in the driveway and a kitchen tent softly lit by kerosene lamps. They go to sleep to the hyenas coyly calling *"ooo-WOOO-ooo"* and wake up to the francolins, plump seed-eating birds, bawling like crows just outside the tent. A hornbill named Hominy lives in camp and steals rice cakes from their young son, Madison.

The dogs McNutt has collared wander through a study area nearly the size of Rhode Island, much of it roadless. He tracks them at times on foot or in a micro-light airplane but mainly by bushwhacking in his Land Rover. He gets three or four thorn-flattened tires a week trying to keep up. When his engine overheats, he cleans the debris out of the radiator screen with a feather from a marabou stork.

Driving out from camp one morning, McNutt picks through the dusty gray tangle of hyena, lion, springhare, and francolin tracks to point out the footprints of wild dogs. "They're very symmetrical, very line-of-direction," he says. "It reflects the balance and light-footedness of the animals as they're moving." He eases down the road, head hanging out the window. "There are three or four dogs here. Cool." He accelerates. "We might just catch up with them."

A few minutes later he spots a dog moving through the woods.

"It's Ditty. She may still be hunting." Two yearlings join her. Their high bellies testify that they have not eaten, but it's time to knock off. They lie down in the shade, undisturbed by McNutt's familiar truck — until I open the side door and take a seat on the ground. This isn't necessarily a bright idea: sitting down in a group of wild animals is the sort of dumb trick that gets bush-macho day-trippers ripped to bits by irritated lions, and quite rightly. Predators deserve more respect than that. But the image of wild dogs as wanton killers is so viscerally embedded in human mythology that it bears firsthand refuting, and McNutt has assured me that I will lose no more than one or two lumps of flesh.

The two yearlings immediately lift their heads. They stand and separate. The sharp edges of their carnassial teeth seem to glint with a scintilla of truth in the old lore. One dog circles behind the truck and begins to creep toward me from the right. The other pads softly through the mopani scrub, head down, and draws closer on my left. A cartoon called "Our Fascinating Earth" leaps to mind, characterizing wild dogs as one of "the most vicious of African carnivores . . . and among the few animals that MAY ATTACK MAN." The dogs advance to within ten feet. Ditty suddenly appears between the two and strides boldly up to the back of my neck. She sniffs once, then drops back, and all three dogs move off, their curiosity satisfied. They flop down in the shade, having deemed me rather a bore.

"These are wild animals," McNutt says when I get back into the truck. "They eat animals the size of us all the time. And they're hungry. And yet they showed no aggression whatsoever." In the course of his fieldwork McNutt has been rammed by an angry hippo, choked with dust when a charging lion skidded to a stop beside his truck, and cornered in the camp shower by a deranged honey badger. But he has never been injured by a wild dog. Ditty's pack has a musical theme, so we name the two yearlings that did not eat me Lyric and Chorus.

In truth, what impresses McNutt about the dogs isn't their viciousness but how gentle and considerate they are with one another. One day we find a pack lying by a great pyramidal termite mound, nose to rump, like any heap of idle, flea-harried house dogs. But every heap has its etiquette: one of the dogs stands, walks ten feet away from the others, sits, and claws furiously at his neck with a hind leg. Then he returns to his place in the heap. A social nicety, McNutt suggests, lest he spread his parasites to a neighbor. Another time, when he had just collared a dog and was waiting for it to regain consciousness, a sibling walked up, grabbed the dog by the collar, and dragged it back to the safety of the pack.

Their highly evolved social etiquette also bears on much larger issues. This heap of dogs, for instance, got its start as a pack in the usual fashion when three brothers from a pack named for mountains joined up with two sisters from a pack named for islands. In most packs only one male and one female do the breeding. The other adults spend their lives helping to rear nieces and nephews. At the moment, in a burrow underneath a termite mound, a female named Cypress (for an island in Puget Sound) is nursing a new litter. She slouches up out of her burrow and approaches one of the other dogs. Dipping her head down under his mouth, she makes a soft mewling sound. His belly begins to heave in response.

By our standards, what follows may sound like an abomination. By theirs, it is selfless everyday caregiving. A nursing female depends on the other dogs to gorge themselves at the kill and then regurgitate back at the den. Some dogs will heave up a portion of their meal seven or eight times a day, especially once the puppies are weaned and begin to beg. The demands of this kind of food sharing are probably the main reason most packs can support only a single breeding female, and here the social etiquette can turn harsh. To reduce competition, McNutt says, the dominant female will often take over or even kill a sister's litter. These aren't our family values, but they are family values nonetheless. "If it's a small pack, maybe five or seven adults," he says, "they're better off having the experienced hunters out hunting, not back at the den rearing young."

The hunt begins one afternoon with the arrival of a small procession of ghouls — hooded vultures, a hyena, and us, all waiting for the wild dogs to go out and kill, a chore for which the dogs themselves appear at first to have no great enthusiasm. They drag themselves up from the heap and mill around, greeting one another. They lean forward and languidly bridge their back legs out behind. One of them moseys off. "This, believe it or not, is it," McNutt says. The others tag along in loose file, with a desultory wagging of tails.

A subordinate named Blackcomb takes the lead, climbing up on a termite mound to peer over the grass. The pace picks up to a trot when there are antelope in sight, then drops back to a walk when their prey escape. McNutt's Land Rover lurches and zigzags to keep up, and the vultures and the hyena hopscotch behind. The harsh glare of midday softens, and the shadows grow longer and more dangerous. The dogs hunt in eerie silence. We hear a single sharp bark — an impala warning its herd — and the dogs instantly jump their pace up to a full run. They are capable of pursuing their prey at twenty-five miles an hour, with bursts up to thirty-five. But when we catch up with a couple of dogs a few minutes later, they are disoriented, seeming to have lost sight of one another and their prey. Blackcomb is absent, so McNutt keeps our truck bumping cross-country in response to whatever he is hearing on his earphones. We find Blackcomb before the other dogs do, with his nose in the warm belly of an impala.

He has made this kill by himself, and the victim's supposedly slow, brutal death appears to have been instantaneous. The impala lies in a single bright patch of blood in the grass. Blackcomb feeds, looks up, feeds again, and finally leaves to bring in his pack mates. The feast that follows takes place, like the hunt, in silence. The dogs grip the carcass from opposite sides, then lift their heads and yank back in unison, as if on the count of three. The only sound is breaking bone and shredding muscle. Cypress, who has come out from the den, twitters softly, and her sister, Gabriola, backs away.

"The thing that distinguishes wild dogs is that they're so easygoing with each other," McNutt remarks. Where wolves would enforce their hierarchy by snarling and showing their teeth, "you can't help but notice how quiet and cooperative the dogs seem to be in the same context." And yet, as Gabriola searches for scraps on the outskirts of the kill, it's clear that a hierarchy is operating here too. Because the subordinate adults are last in line at the kill, says McNutt, they'll be more motivated to make a kill next time. The risk of leading the hunt brings a subordinate the reward of cramming its belly for a few minutes, as Blackcomb did, before the twinge of social conscience causes it to bring in the rest of the pack. "It's a neat system."

On the way back to camp, McNutt speculates on why wild dogs have evolved into such thoroughly social creatures. They typically

travel in packs of about ten individuals, in part because group living comes at relatively little cost. The most one dog, weighing forty
to eighty pounds, can stuff down its gut at a feeding is about ten
pounds of meat, but their prey average more than one hundred
pounds. So food for one is food for a crowd. Each extra mouth also
brings a pair of those acutely sensitive satellite-dish ears for added
vigilance against "kleptoparasites" like hyenas, which might easily
steal the kill of a solitary dog. The pack also provides protection
against the bane of wild dog life, which is lions.

One evening when Blackcomb is again leading the hunt, he suddenly stops for no visible reason and rears up on his hind legs. His
brother Tremblant joins him, peering a hundred yards ahead and
making a low, rolling *ru-ru-ru* growl, which means "There's a lion
out there." The lion, a subadult, yawns massively in the face of the
dogs. "It's that old cat-and-dog thing," McNutt remarks. The lion
eventually gets up and plods off into a field of phragmites, the
feathery seed heads backlit by the setting sun so they flame like a
thousand torches. Blackcomb and the others follow, close enough
to nip at the lion's haunches. The lion spins on them and snaps but
continues his retreat. Then two more lions appear, and the dogs
suddenly recall, with a parting *ru-ru-ru,* that they had an appoint-
ment with an impala on the far side of town. In one study, predators — almost always lions — killed 42 percent of the wild dog juve-
niles and 22 percent of the adults. Humans are the other great
cause of wild dog mortality, and these two factors, combined with
the footloose behavior of the species, are the reason wild dogs present such a challenge for conservation.

Except during the denning season, wild dogs seldom stay in one
place for more than a day or two. In the Okavango a typical pack
wanders through a home range of about 175 square miles, and
nearly four times that in the Serengeti. Few national parks in Africa
are big enough to sustain a healthy population of wild dogs. And in
almost every national park the dominant species — and the most
popular tourist attraction — is the lion. If the dogs seek refuge
from lions by going outside the parks, they quickly come into conflict with humans, usually after they kill one of the cattle that have
displaced their traditional prey.

Thus almost every attempt over the past twenty years to repopulate parks with wild dogs has failed dismally: "Starved or killed by lions within 4 months. . . . Shot on nearby farm. . . . Left the reserve and were poisoned." An exception is at Madikwe Game Reserve, a day's drive from the Okavango, on South Africa's border with Botswana. Madikwe is an experiment, an artificial park of about 230 square miles created over the past eight years on derelict ranchland, primarily to bring tourist revenue into South Africa's North West Province and only secondarily for conservation. It's enclosed by a fence more than ninety miles long, built with steel reinforcing cable and a 7,000-volt electric wire. The creators of the park have established a balanced population of prey and predators, including just enough lions to gratify tourists. But they have chosen to make wild dogs a featured attraction. One Madikwe staffer puts it this way: "Lions are common as muck in South Africa. Wild dogs are not."

Madikwe now has two packs of dogs, put together as a sort of blind date between wild-caught males from Botswana and South Africa and captive-bred females. One pack has already produced its first litter. Some conservationists hope eventually to establish wild dogs in a half dozen new Madikwes around southern Africa, including new transborder national parks and private conservation areas formed by neighboring game ranches. To maintain genetic diversity, these parks would swap breeding stock, much as zoos do now. What Madikwe promises is a future in modern Africa for wild dogs — if only as a managed, marketed, and fenced-in species.

This is an approach Tico McNutt finds deeply, almost inexpressibly disturbing. "I don't believe we're going to get very far," he says carefully one evening when we are out watching dogs, "if we justify conservation only by assigning an economic value to an animal or an ecosystem. Surely that's not the only reason. It's not the reason I'm interested in conservation."

Even the creators of Madikwe argue that it would be better to preserve existing wilderness than to attempt to re-create it. But they also say that economic values are what actually motivate people to save a wild area in the first place. "You don't realize its value until it's gone, and then it takes an enormous amount of capital to reestablish it," says Richard Davies, project manager for Madikwe. His business plan is for Madikwe to generate about $17 million a year in revenue by 2010, with some of the profits going to neighboring communities. "There's a strong lesson here for countries to the north of us that are squandering their wildlife," says Davies, and he means Botswana in particular.

This is the dispiriting subtext to Tico McNutt's research: He names his wild dogs, records their genealogies, and chronicles their footloose lives in the expectation that they will not be able to live this way much longer, even in a wilderness as vast as the Okavango. On the surface it's a familiar story of villagers with cattle steadily encroaching on wildlife. In Shorobe, a cluster of mud huts on the edge of the Okavango, a group of threadbare farmers sit in the dust to talk beside a well they have just drilled for a new water hole, and they sound like livestock ranchers everywhere. They long to kill predators. They gripe about government compensation programs, which pay for their lost animals slowly or not at all. They live outside the loose perimeter known as the southern Buffalo Fence, and they occasionally lose livestock to roaming lions, hyenas, and wild dogs.

When I suggest it might be better not to keep cattle this close to a wildlife refuge, my translator does me a favor by refusing to translate, and explains: "Wildlife belongs to the government, and livestock belongs to the farmer. If it gets into the farmers' minds that you think wildlife is more valuable than cattle, then you will be starting a fire." And he adds, "According to Botswana culture, you cannot live without cattle." It could be Wyoming, outside Yellowstone National Park. But the farmers are also attuned to new possibilities. They envy another village up the road that operates its district as a wildlife management area and profits from concessions for hunting, sightseeing, and tourist lodges. "We can live with wildlife on one side of the Buffalo Fence and livestock on the other," a farmer says. "All we want," says another, "is some benefit from our natural resources."

Environmental critics say the larger threat to the Okavango and its wild dogs is commercial cattle ranching. This industry is heavily subsidized by the European Union, which opened its markets to Botswana beef in 1972 on condition that the cattle come from disease-free areas. To limit disease, the country built a network of veterinary cordon fences. These fences have cut off ancient animal migration routes. In one notorious incident in 1983, 50,000 wildebeests piled up dead against a new fence that prevented them from reaching water. Populations of some species have plummeted by more than 80 percent just since 1978, and the fences have lately begun to close in around the last great enclave of wildlife in the Okavango.

"We're seeing significant loss of range for large mammals, and as their numbers go down, predators must also go down," says Karen Ross, director of Conservation International's Okavango Program. She counts more than nine hundred miles of new fences erected around the delta just since 1995. "If cattle move into this area, conflicts with wild dogs are going to increase."

According to a recent report from the University of Botswana, the economic benefits of the European subsidy program have gone almost exclusively to commercial ranches controlled by the nation's wealthy ruling elite, not to rural villagers. The same powerful interests are likely to benefit if the Okavango floodplains are converted to ranchland. "People aren't starving in Botswana," Tico McNutt says. "It has to do with a small number of people getting an economic gain out of it."

Out in the Okavango one evening, McNutt and I are talking about dogs and doing our best not to think about all that. Probably we should be savoring all that is sublime and unfettered about wild dogs in their natural element. But the truth is that at the moment we are just having a good time. We joke about the sly twist of destiny that caused three brothers from the Painters pack (Braque, Bacon, and Rothko) to hook up for a time with the Four Females from Hell before settling down with a trio of females named for single-malt whiskies (Tamdhu, Islay, and Talisker). I suggest that following the different packs as their lives unfold over time must have a quality like soap opera for him.

"The thing that motivates me most to stay on a set of tracks all day and again the next day," he says, "is to find out if it's one of the hundreds of dogs I've come to know. You've been with their mothers and fathers when they were born, and you see them grow to reproductive age and then disperse and disappear. When you find them again, it's exciting."

"So tell me about the Four Females from Hell," I say. Their names, he says, were Trumpet, Viola, Tympany, and Bell, and at various times he saw them with seven different groups of males, several of which died or disappeared soon afterward. Among the suitors, somewhere between the Painters and Toto from the Wizard of Oz pack, was a male named Piccolo. "Then the females disappeared," McNutt says, and after two years he figured that lions or farmers had killed them. But one day a new pack showed up in his study area, and it dawned on McNutt that the male was Piccolo and the female was Bell. "I found them hanging out together with yearlings," he says. They had become the parents of Ditty, the same dog that sniffed at the back of my neck, and Lyric and Chorus, who did not eat me. "At least one of the Four Females from Hell had successfully reproduced and stabilized," McNutt says, gratified, and with an "I knew the bride when she used to rock-and-roll" sort of smile.

Later a solo male showed up on the fringes of the pack, and it turned out that McNutt knew him too. His name was Newkie (short for Newcastle), a Beer-pack dog whose older brother had once courted the Four Females from Hell. Newkie also had a history in McNutt's notebooks. McNutt had watched him grow to reproductive age and then strike out on his own. While he was away, an epidemic hit the Beer pack, possibly rabies or canine distemper picked up from a villager's dog. "Newkie came back and found everyone dead," McNutt says. Newkie settled into his old home, finding solace for his social nature in the scent marks of his pack, which lingered like ghosts for months afterward. Then he began to shadow the Music pack, hoping to lure away Ditty or possibly to replace Piccolo as the dominant male. Ditty showed no interest, and Piccolo repeatedly pushed him off. But McNutt noticed that Piccolo's rebellious son Riff sometimes ran interference for Newkie against Piccolo. Now Riff had left his home pack to join up with Newkie. Together, says McNutt, they have a better chance of attracting females than either of them would have on his own.

But this is as far as the story goes this evening. McNutt cannot say if Newkie will get a girl, or if Ditty will finally strike out on her own, or if Bell and Piccolo will grow old and dowdy together. He will have to stay tuned for the next episode. We watch the birds known as queleas come rolling into their evening roosts, undulating like swarms of insects above the marsh. A couple of red-necked falcons pick off stray birds to eat for dinner. In the distance a hippo sounds its sonorous bassoon note.

The Land Rover turns back to camp, and it occurs to me that the lives of the dogs are as messy and tangled as our own. As rich with the tidal coming and going of generations. McNutt is wheeling around trees and stray elephants, muttering, "Uh-oh, uh-oh," and I am thinking about another night when I watched a litter of yearlings playing in the dark. A half dozen of them chased each other in a tight circle around a sage bush, diving into the middle, then shooting out the sides. They jaw-wrestled and played tug-of-war with one another. They made mock charges and danced apart, then stood with mouths slightly open, eyes bright, seeming to grin. One paused to catch his breath, as if having called time-out. Then he crept up to pounce on a littermate and set the chase going again. All this took place, like almost everything wild dogs do, in silence. The sounds we heard in the darkness were the dry rustling of the bush, the huffing of the dogs' breaths, the soft, horselike thumping of their footpads on dry earth. A person passing by fifty feet away might have thought there was nothing much going on out there.

We left them like that, dancing together in the darkness. "You always leave wanting to come back," a friend had told me. With luck the dogs might still be there if I ever get the chance. They would be lying in their doggy heaps or hunting like dappled shadows at dusk. Better to think about that, I figured, than the other possibility, which is that soon there may be no wild dogs at all.

PAUL DE PALMA

http://www.when_is_enough_enough?.com

FROM *The American Scholar*

IN THE MISTY PAST, before Bill Gates joined the company of
the world's richest men, before the mass-marketed personal computer, before the metaphor of an information superhighway had
been worn down to a cliché, I heard Roger Schank interviewed
on National Public Radio. Then a computer science professor at
Yale, Schank was already well known in artificial intelligence circles. Because those circles did not include me, a new programmer
at Sperry Univac, I hadn't heard of him. Though I've forgotten the
details of the conversation, I have never forgotten Schank's insis-
tence that most people do not need to own computers.

That view, of course, has not prevailed. Either we own a personal
computer and fret about upgrades, or we are scheming to own one
and fret about the technical marvel yet to come that will render
our purchase obsolete. Well, there are worse ways to spend money,
I suppose. For all I know, even Schank owns a personal computer.
They're fiendishly clever machines, after all, and they've helped
keep the wolf from my door for a long time.

It is not the personal computer itself that I object to. What
reasonable person would voluntarily go back to a typewriter? The
mischief is not in the computer itself, but in the ideology that sur-
rounds it. If we hope to employ computers for tasks more interesting than word processing, we must devote some attention to how
they are actually being used, and beyond that, to the remarkable
grip that the idol of computing continues to exert.

A distressing aspect of the media attention paid to the glories of
technology is the persistent misidentification of the computing sci-

ences with microcomputer gadgetry. This manifests itself in many ways. Once my seatmate on a plane learns that I am a computer science professor, I'm expected to chat about the glories of the new DVD-ROM as opposed to the older CD-ROM drives; or about that home shopping channel for the computer literate, the World Wide Web; or about one of the thousand other dreary topics that fill *PC Magazine* and your daily paper, and that by and large represent computing to most Americans. On a somewhat more pernicious level, we in computer science must contend with the phenomenon of prospective employers who ask for expertise in this or that proprietary product. This has had the effect of skewing our mission in the eyes of students majoring in our field. I recently saw a student résumé that listed skill with Harvard Graphics but neglected to mention course work in data communications. Another recent graduate in computer science insisted that the ability to write WordPerfect macros belonged on her résumé.

This is a sorry state. How we got there deserves some consideration.

A few words of self-disclosure may be in order. What I have to say may strike some as churlish ingratitude to an industry that has provided me with a life of comparative ease for nearly two decades. The fact is that my career as a computer scientist was foisted upon me. When I discovered computers, I was working on a doctorate in English at Berkeley and contemplating a life not of ease but of almost certain underemployment. The computer industry found me one morning on its doorstep, wrapped me in its generous embrace, and has cared for me ever since. I am paid well to puzzle out the charming intricacies of computer programs with bright, attentive students, all happy in the knowledge that their skills will be avidly sought out the day after graduation. I can go to sleep confident that were tenure to be abolished tomorrow, the industry would welcome me back like a prodigal son.

Yet for all its largesse, I fear the computer industry has never had my full loyalty. Neither did English studies, for that matter, but this probably says more about those drawn to the study of texts than about me. My memories of the time I spent in the company of the "best which has been thought and said" are hazy, perhaps because the study of literature is not so much a discipline as an attitude.

The attitude that dominated all others when I was a student, that sustained my forays into the Western Americana of the Bancroft Collection, is that there is no text so dreary, so impoverished, so bereft of ideas that it does not cry out to be examined — deconstructed, as a graduate student a few years my junior might have said. But the text I now propose to examine, impelled, as it were, by early imprinting in the English department, goes beyond words on a page.

From an article here and a TV program there, from a thousand conversations on commuter trains and over lunch and dinner, from the desperate scrambling of local politicians after software companies, the notion that prosperity follows computing, like the rain that was once thought to follow the settler's plow, has become a fully formed mythology.

In his perceptive little book *Technopoly,* Neil Postman argues that all disciplines ought to be taught as if they were history. That way, students "can begin to understand, as they now do not, that knowledge is not a fixed thing but a stage in human development, with a past and a future." I wish I'd said that first. If all knowledge has a past — and computer technology is surely a special kind of knowledge — then all knowledge is contingent. The technical landscape is not an engineering necessity. It might be other than it is. Our prospective majors might come to us, as new mathematics or physics majors come to their professors, because of an especially inspiring high school teacher, because of a flair for symbol manipulation, or even because of a (dare I use the word?) curiosity about what constitutes the discipline and its objects of study — not simply because they like gadgets and there's a ton of money to be made in computing.

The misidentification of computer science with microcomputer gadgetry is a symptom of a problem that goes far beyond academe. Extraordinary assertions are being made about computers in general and microcomputers in particular. These assertions translate into claims on the American purse — either directly, or indirectly through the tax system. Every dollar our school districts spend on microcomputers is a dollar not spent reducing class size, buying books for the library, reinstating art programs, hiring school counselors, and so on. In fact, every dollar that each of us spends outfitting ourselves with the year's biggest, fastest microcomputer is a dollar we might have put away for retirement, saved for our children's education, spent touring the splendors of the American

West, or even chosen not to earn. In the spirit of Neil Postman, then, I'd like to speculate about how the mythology of prosperity through computing has come to be and, in the process, suggest that like the Wizard of Oz, it may be less miraculous than it looks.

The place to begin is the spectacular spread of microcomputers themselves. By 1993 nearly a quarter of American households owned at least one. Four years later, the *Wall Street Journal* put this figure at over 40 percent. For a home appliance that costs at least $1,000, probably closer to $2,000, this represents a substantial outlay. The home market, as it turns out, is the smaller part of the story by far. The Census Bureau tells us that in 1995, the last year for which data are available, Americans spent almost $48 billion on small computers for their homes and businesses. This figure excludes software, peripherals, and services purchased after the new machines were installed.

The title of an article in the *Economist* — "Personal Computers: The End of Good Times?" — hints at the extraordinary world we are trying to understand. In it we learn that annual growth in the home computer market slowed from 40 percent in 1994 to between 15 percent and 20 percent in 1995. By the fall of 1998, market analysts were predicting 16 percent growth in the industry as a whole for the current year. Those of us involved in other sectors of the economy can only look on in astonishment. When a 20 percent, or even 16 percent, growth rate — well over five times that of the economy as a whole — is "the end of good times," we know we're in the presence of an industry whose expectations and promises have left the earth's gravitational pull.

To put some flesh on these numbers, let's try a thought experiment. The computer on my desk is about 16 inches by 17 inches. The Census Bureau tells us that the microcomputer industry delivered over 18 million machines in 1994, the year when, according to the *Economist,* good times ended. Of these, perhaps a third went to the home market, the balance to business. At the 40 percent growth rate in the home market cited for that year and the more modest 16 percent growth rate for the business market, the boys in Redmond and Silicon Valley will have covered the United States' 3,679,192 square miles with discarded microcomputers well before my daughter, who is now thirteen, begins to collect Social Security.

Fabulous as they seem, these figures come from only part of the industry. Microcomputers do not define computing, despite their spectacular entry on the scene. The standard story goes like this: There was once a lumbering blue dinosaur called IBM that dominated the computer industry. In due course, smaller, more agile, and immensely more clever mammals appeared on the scene. The most agile and clever of these was Microsoft, which proceeded to expand its ecological niche and, in so doing, drove the feebleminded IBM to the brink of extinction.

The business history in this story is as faulty as its paleontology. IBM may be lumbering and blue, but in 1997 its sales were nearly $78 billion. Compare that with Microsoft's $9 billion. The real story is not in the sales volumes of the two companies but in their profit margins. In 1997 IBM's was 7.7 percent, while Microsoft's was a spectacular 28.7 percent. This almost mythical earning capability is expressed best in *Forbes*'s annual list of very rich Americans. We don't hear much about IBM billionaires these days, but Microsoft fortunes are conspicuous in the *Forbes* list, with Bill Gates's $51 billion, Paul Allen's $21 billion, and Steven Ballmer's $10.7 billion. These fortunes were accumulated in less than twenty years from manufacturing a product that requires no materials beyond the inexpensive medium it is stored on — not so different from a pickle producer, whose only cost, after the first jar comes off the line, is the jar itself. It's a tale of alchemical transmutation if ever there was one. Is it really a surprise that most people don't know that IBM is still a very successful company or that computer science does not begin and end with Windows 98?

This joyous account of fortunes waiting to be made in the microcomputer industry has a dark side. Just as Satan is the strongest character in *Paradise Lost*, as C. S. Lewis observed, so is popular fascination with computers due as much to the dark side as to the light. Despite generally good economic news for the past few years, Americans seem gloomy about their prospects. Our brave new world, paved over with networked computers from sea to shining sea, may well be one in which we are mostly unemployed or have experienced a serious decline in living standards. Computers, if not always at the center of the problem, are popularly thought to have been a major contributing factor.

Look at the substantial decline in manufacturing as a segment of the workforce in the United States. Between 1970 and 1996 (the last year for which data are available), the number of Americans employed increased by about 50 million. During this same period, the number of manufacturing jobs declined by about 200,000. The culprit here is often thought to be computer technology, through assembly line robots or through U.S.-owned (or U.S.-contracted) manufacturing facilities in developing countries. Asia and Latin America, of course, would have less appeal to American corporations without worldwide data communications networks.

This analysis of the decline in manufacturing employment is perhaps more appealing than true. I will return to the relationship between computers and productivity. For now it's enough to observe that most people believe there is such a relationship. So if the money to be made in the computer industry is not sufficient inducement to vote for the next school bond issue that would outfit every classroom in your city with networked computers, then the poverty your children will certainly face without such a network should do the trick. With those staggering Microsoft fortunes in the background and the threat of corporate retrenchment in the foreground, I suppose I'm naive to expect the strangers I chat with on planes to know that the computing sciences are more like mathematics and the physical sciences than like desktop publishing — or, for that matter, like the rush to the Klondike goldfields.

The emergence of the microcomputer as a consumer item in the past decade and a half has prompted a flood of articles in the educational literature promoting what has come to be called "computer literacy." In its most basic sense, this term appears to refer to something like a passing familiarity with microcomputers and their commercial applications, rather like the ability to drive a car and know when to get the oil changed. Sadly, the proponents of computer literacy have won the high ground by virtue of the term itself. Who would argue with literacy? It is, after all, one of the more complex human achievements. Not only is literacy a shorthand measure of a country's economic development, but as the rhetorician Walter J. Ong has long argued, once a culture becomes generally literate, its modes of conceptualization are radically altered. Literacy — like motherhood and apple pie in the America of my youth — is unassailable.

But what about the transformative nature of literacy? I am fully

aware that similar claims have been made about computers —
namely, that computers, like writing, will alter our modes of conceptualization. Maybe so, but not just by running Microsoft Office.
I've developed a rule of thumb about claims of this sort: If the
subject matter is computers and the tense of the claim is future
(and, therefore, its truth-value cannot be ascertained), look at the
subtext. Is the claimant a salesman in disguise? To recognize the
nonsense in the claim that computers will transform the way we
think, we need only indulge in some honest self-examination. I
would give up my word processor with great reluctance. This
doesn't mean that my neuronal structure is somehow fundamentally different from what it was when I was writing essays similar to
this one on my manual Smith Corona. It does mean that the computer industry is a smidgen richer because of my contribution. It
also means, as was recently pointed out to me, that it is a good bit
easier to run on at great length on a computer than on a typewriter.

Not surprisingly, the number of articles addressing computer literacy in the educational literature has kept pace with microcomputer developments. ERIC is a database of titles published in education journals. When I searched ERIC using the keywords *computer
literacy* and *computer literate,* I found 97 articles for the years 1966–
1981, or an average of about 6 per year. The decade from 1982 to
1991 produced 2,703 hits, or about 270 per year. At first glance the
production of articles since 1991 shows welcome signs of dropping off. But the Internet has come to the rescue of both the microcomputer industry and its prognosticators. When I add the terms
"Internet," "World Wide Web," and "information superhighway" to
the mix (subtracting for duplicates), the total rises to an astonishing 4,680 articles from 1992 through the first half of 1998. This
works out to about 720 articles per year. The bulk of the recent articles, of course, are full of blather about the so-called information
superhighway and how all those school districts that cannot give
every child access to it will be condemning the next generation to
lives of poverty and ignorance.

Since computer literacy advocates are eloquent on the benefits
of computers in our schools (and equally eloquent on the grim fate
that awaits those students not so blessed), a brief look at how microcomputers are actually used in primary and secondary schools
is in order. Microcomputers are now a solid presence in American

education. The U.S. Census Bureau put the number at nearly 7 million in 1997, or just over 7 students per machine, compared with 11 students per machine in 1994 and 63 per machine a decade earlier. Picture a classroom richly endowed with computers. Several students are bent over a machine, eyes aglow with the discoveries unfolding on the screen. Perhaps there is a kindly teacher in the portrait, pointing to some complex relationship that the computer has helped the budding physicists, social scientists, or software engineers to uncover. If this is the way you imagine primary and secondary school students using computers, you are dead wrong. Several important studies have concluded that primary and secondary school students spend more time mastering the intricacies of word processing than they do using computers for the kinds of tasks that we have in mind when we vote for a bond issue.

Programming, in fact, was the one area that school computer coordinators saw decline over previous years. I would be the first to acknowledge that programming does not define computer science. This simple fact is what makes the endless discussion of programming languages in computer science circles so tedious. Nevertheless, if computer science does not begin and end with programming, neither will it give up its secrets to those who cannot program. I greet the news that high school students do not program our millions of microcomputers as an English professor might greet the news that the school library is terrific but the kids don't read. Here is a puzzle worth more than a moment's thought. There is an inverse relationship between the availability of microcomputers to primary and secondary school students and the chance that those students will do something substantial with them. I am not saying that the relationship is causal, but the association is there. Draw your own conclusions.

Though the jury is still out on the potential educational benefits of computing, we all agree that skill with computers is necessary for success in business. Even here there's a problem. Recent studies have assembled evidence that should give computer enthusiasts some sleepless nights. It appears that most businesses would be better off had they taken all that money they spent on computer technology and put it into bonds at market rates. This investment has been substantial, as anyone knows who has seen the piles of unopened software, the manuals still in their shrink-wrapped plastic, and the stacks of obsolete hardware accumulating in storerooms around the country. By 1995 it had totaled over $4 trillion. This sum, it should be noted, excludes the public money involved in training (and employing) academic computer scientists and engineers.

It also excludes another hidden expenditure. The time employees spend rearranging icons on their screens, the time they spend wondering why their spreadsheets will not recognize their printers — in fact, all those minutes here and hours there spent fiddling with hardware and software — is time they do not spend on the tasks they are being paid to perform.

Let me tell a story. I have been a computer science professor for seven years. Before that I spent a decade working for some of the largest firms in the computer industry. I am, by any reasonable measure, computer literate. One recent Sunday afternoon I thought I might pop into my office, copy this essay to a floppy disk, and work on it at home, where I was also caring for a child with chicken pox. In other words, I took the microcomputer industry up on its central promise: workers will be liberated from the tyranny of place. Able to function as parents and workers simultaneously, we will prosper along with our employers.

Here's what really happened. I promised my wife I'd be gone no more than thirty-five minutes, twenty-five for the drive to and from the university, ten to copy the file. As it happens, I am the last remaining member of the professional middle class without Microsoft Windows running at home. What I have is a 286 IBM clone running DOS and WordPerfect 5.1, equipment my last employer gave me when I left seven years ago. This setup was well on its way toward obsolescence when I acquired it. Were my students to learn that I wrote with a quill pen by the light of an oil lamp, they would think me hardly less quaint.

However, my reluctance to part with hard-earned money for a shiny new computer that I would use only as an abundantly outfitted typewriter did pose a small challenge. I would have to get the file from my office computer, a fancy Pentium workstation (courtesy of my current employer), to run on my ever-faithful home machine. Not a problem, I thought. I can easily transform the Microsoft Word file in my office to ASCII text, copy it to a 5¼-inch floppy disk, and read it into WordPerfect at home.

As we have all come to know, painfully at first and finally with res-
ignation, when the subject is personal computers, things are not always as promised. (It has occurred to me more than once that the
computer industry should have the honor of Iago and display these
words boldly on every screen: "I am not what I am.") First I learned
that my document was infected with the Word macro virus. No
matter how I tried, Word would not let me transform it from a tem-
plate (a term known to all Word users, happily ignored by most) to
a text file. So I called a colleague who gave me what is known in the
computer industry as a "workaround." A workaround is what you
do when your machine is not running the way the manufacturer
promised. By analogy, a workaround for faulty automobile brakes
might be to open the door and drag your feet. In any case, my colleague is a clever fellow and the workaround did allow me to work
around the handiwork of the disgruntled Microsoft employee who
had infected Word with the virus. Having transformed my essay
into a generic text file, I was ready to copy it to a floppy and return
home, safe in the knowledge that I could be both productive and
parental.

Unfortunately, our former systems administrator, for reasons
that must have made eminent sense to him, had disabled my A
drive. But as I said earlier, I am computer literate. Though annoy-
ing, a disabled drive is not catastrophic. I need only invoke a special setup routine to let Windows knows that, in fact, there is a $5\frac{1}{4}$-
inch floppy drive on my machine. But since this machine is a
castoff from our department's lab, the systems administrator had,
wisely, password-protected the setup routine. He had also, in
the meantime, decamped for the vastly more remunerative pastures of the computer industry. In a word, he was unavailable and
so was my machine. I returned home, nearly two hours after I
had left, to an unhappy wife and a sicker child — and without
the file.

This is not an isolated story. Anyone who has dealt with a microcomputer has a store of similar tales. There is another story here
as well. Even if one is inclined to stick with the tried and true,
the computer industry — and its minions across the land — will
not permit it. By the time this article goes to press, the last computer in my department with a drive that accommodates $5\frac{1}{4}$-inch
diskettes will have gone to wherever old computers go. My well-
worn and well-loved 286 will then be an island cut off from the

main, and I, its single inhabitant, will speak a language that is fast
becoming extinct.

The price of computing equipment has dropped dramatically in
recent years. For under $2,000 you can buy a microcomputer that
processes millions of instructions per second and is equipped with
a stunningly large memory and disk space. At that price we can
all be equipped at the office, and most of us will choose to be
equipped at home. As it happens, that $2,000 (plus a bit more
for networking components) is the smallest slice of the great pie of
microcomputer costs. The *Economist* cites a study by the Gartner
Group, a respected consulting firm, that puts the annual cost of a
microcomputer connected to a network at $13,200. Of this, only
21 percent goes to the purchase of hardware and software. Administrative costs absorb 36 percent. We have to pay all those people
who come to our rescue, after all. This figure alone should slow
down the headlong rush to outfit every desk on the planet with a
microcomputer.

That administration costs more than the machine itself is not the
biggest surprise. Recall my story. Just how much was my two hours
worth? On the average, 43 percent of the cost of a microcomputer
is consumed in what Gartner calls "end-user operations." Just what
are end-user operations? They are all those things one does with a
computer in order to do the things one gets paid to do. This includes rearranging icons, coaxing disk drives into action, loading
and setting up software, avoiding viruses, listening to Microsoft's
music as you wait helplessly on hold for advice from someone in
technical support who probably knows less than you do, and so on.

Though the Gartner Group has done us the service of quantify-
ing those long hours spent mastering yet another Microsoft user in-
terface, the effect of that time on worker productivity has been
known for some years now. Many studies, including some done by
the National Research Council and by Morgan Stanley, the New
York investment bank, fail to indicate any correlation between productivity growth and information technology expenditures. Dis-
tressingly, the opposite appears to be true. As Thomas Landauer
has pointed out in *The Trouble with Computers,* those industries that
invested most heavily in information technology, with the exception of communications, seem to have the most sluggish productiv-

ity growth rates. Though one still could argue that schools and colleges should continue to teach courses in microcomputer literacy because microcomputer usage has grown like a fungus after a heavy rain, our time might be more profitably spent breaking the bad news to the public that pays the bills. In the process, we might also come to understand how a machine so patently clever as the microcomputer could have done the business world (outside of the computer industry itself) so little good.

Given the several thousand articles on computer literacy and the emerging inverse relationship between productivity growth and computer expenditures, it seems reasonable to ask just who does benefit from the computer literacy movement — and who pays for it. The commonsense answer is, Students benefit. Well, common sense is right, but, as usual, only partially so. Students, of course, are served by learning how to use microcomputers. But the main beneficiaries are the major producers of hardware and software. The situation is really quite extraordinary. Schools and colleges across the country are offering academic credit to students who master the basics of sophisticated consumer products. Granted that it is more difficult to master Microsoft Office than it is to learn to use a VCR or a toaster oven, the difference is one of degree, not of kind.

The obvious question is why the computer industry itself does not train its customers. The answer is that it doesn't have to. Schools, at great public expense, provide this service to the computer industry free of charge. Not only do the educational institutions provide the trainers and the setting for the training, they actually purchase the products on which the students are to be trained from the corporations that are the primary beneficiaries of that training. The story is an old but generally unrecognized one in the United States: the costs are socialized, while the benefits are privatized.

I have described a bleak landscape in this essay. Let me summarize my observations:

Schools and universities purchase products from the computer industry to offer training that benefits the computer industry.

These purchases are both publicly subsidized through tax support and paid for by students (and their parents) themselves.

The skill imparted is at best trivial and does not require faculty with advanced degrees in computer science — degrees acquired by and large through public, not computer industry, support.

As the number of microcomputers in our schools has grown, the chance that something interesting might be done with them has decreased.

The stunning complexity of microcomputer hardware and software has had the disastrous effect of transforming every English professor, every secretary, every engineer, every manager into a computer systems technician.

For all the public subsidies involved in the computer literacy movement, the evidence that microcomputers have made good on their central promise — increased productivity — is, at the very least, open to question.

If my argument is at least partially correct, we should begin to rethink computing. The microcomputer industry has been with us for a decade and a half. We have poured staggering sums down its insatiable maw. It is time to face an unpleasant fact: the so-called microcomputer revolution has cost much more than it has returned. One problem is that microcomputers are vastly more complex than the tasks ordinarily asked of them. To write a report on a machine with a Pentium II processor, sixty-four megabytes of memory, and an eight-gigabyte hard disk is like leasing the space shuttle to fly from New York to Boston to catch a Celtics game. Though there are those who wouldn't hesitate to do such a thing if they could afford it (or get it subsidized, which is more to the point), we follow their lead at great peril. The computer industry itself is beginning to recognize the foolishness of placing such computing power on every office worker's desk. Oracle, the world's premier manufacturer of database management systems; Sun Microsystems, a maker of powerful and highly respected engineering workstations; and IBM itself are arguing that a substantially scaled-down network computer, costing under $1,000, would serve corporate users better than the monsters necessary to run Microsoft's products.

Please don't misunderstand. This is not a neo-Luddite plea to toss computers out the window. I am, after all, a computer science professor, and I am certainly not ready (as the militias in my part of the country put it) to get off the grid. Further, the social benefits of computing — from telecommunications to business transactions to medicine to science — are well known. This essay is simply a plea to think reasonably about these machines, to recognize the hucksterism in the hysterical cries for computer literacy, to steel ourselves against the urge to keep throwing money at Redmond and Silicon Valley.

Putting microcomputers in their place will also have a salutary effect on my discipline. We in computer science could then begin to claim that our field — like mathematics, like English literature, like philosophy — is a marvelous human creation whose study is its own reward. To study computer science calls for concentration, discipline, even some amount of deferred gratification, but it requires neither Windows 98, nor a four-hundred-megahertz Pentium II processor, nor a graphical Web browser. Though I am tempted, I will not go so far as to say that the introductory study of computer science requires no computing equipment at all (though Alan Turing did do some pretty impressive work without a microcomputer budget). We do seem, however, to have confused the violin with the concerto, the pencil with the theorem, and the dancer with the dance.

I am afraid that we in computing have made a Faustian bargain. In exchange for riches, we are condemned to a lifetime of conversations about the World Wide Web. An eternity in hell with Dr. Faustus, suffering the torments of demons, would be an afternoon in the park by comparison.

HELEN EPSTEIN

Something Happened

FROM *The New York Review of Books*

1.

IN THE MOVIE VERSION of H. G. Wells's *The Island of Dr. Moreau,*
a shipwrecked passenger is fished out of the sea by a cargo ship
on its way to deliver crates of apes, lions, and other wild animals
to a reclusive and mysterious scientist on an island in the South
Pacific. At one point the passenger asks a member of the ship's
crew, "Hey, what is all this mystery about Moreau and his island?"
"I don't know," says the sailor. "If I did know, maybe I'd want to
forget."

Dr. Moreau is tinkering with evolution. He has moved to this remote island laboratory to create a hybrid race of beast-men. He
wants to change "the physiology, the chemical rhythm" of these
creatures, something similar, he says, to vaccination. Moreau believes he is working in the service of mankind and science, but in
the end his beast-men turn on him and destroy his laboratory and
everything else on the island. Reading *The River,* Edward Hooper's
book about the origin of the AIDS epidemic, I began to wonder
whether Hooper had ever read Wells's book or seen *Island of Lost
Souls.* I also wondered just how prescient Wells might have been
about what goes on in modern biology labs. "The spirit of Dr.
Moreau is alive and well and living in these United States," says the
science-fiction writer Brian Aldiss in the afterword to a recent edition of Wells's book. "These days, he would be state funded."

Thirty-three million people living in the world today carry HIV.
So far 14 million have died of AIDS. In some urban areas in Bot-

swana, Zimbabwe, and Malawi, more than 20 percent of sexually active adults are infected with HIV. The UN estimates that AIDS is responsible for 5,500 funerals a day in Africa. The disease is ruining families, villages, businesses, and armies and leaving behind a sadness so immense that it may take hundreds of years to heal.

AIDS is caused by a family of viruses called HIV that destroy the immune system that protects the body from disease. These viruses pass from person to person in bodily secretions during vaginal, anal, or oral sex; they also pass through blood transfusions and pharmaceuticals made from blood products and through bloody hypodermic needles. Infected mothers may also transmit these viruses to their babies in the womb or through breast-feeding. There is no cure, and the immune defenses of even the lucky patients who can afford the newest, most expensive treatments eventually falter. Death, often painful, follows diseases caused by strange microbes that once were extremely rare.

Some African monkeys and apes carry viruses that closely resemble HIV, and most AIDS researchers now believe that the HIV viruses are really primate viruses that somehow jumped into human beings. HIV-1, the virus responsible for most cases of AIDS to date, probably came from a chimpanzee, and HIV-2, a less aggressive virus more common among West Africans, almost certainly came from a monkey called the sooty mangabey.

In thousands of years of human history, not a single person is known to have been infected with HIV until 1959. It is possible that there were cases before that year, but no colonial medical officer, African doctor, or traditional healer ever noticed them. Now 16,000 people become infected with HIV every day and 7,000 people die of AIDS. Why have these horrible viruses started killing people in such vast numbers now and never before? Numerous explanations have been proposed. AIDS has been said to be a divine act, punishing humanity for its godless ways. Or the viruses were allegedly concocted in germ warfare labs in Maryland or the USSR or at the UN and then escaped. It has even been claimed that they came from outer space. Hooper has devoted more than a decade to this question, and he does not seem to know for certain why HIV turned up now and never before. His best guess, however, is that

HIV emerged from an accident in a jungle laboratory in the 1950s, not all that much unlike Dr. Moreau's.

In the 1950s a group of scientists associated with the American pharmaceutical company Lederle, and later the Wistar Institute in Philadelphia, were racing to develop the first polio vaccine that could be given by mouth instead of by injection. Hilary Koprowski, a Polish-American vaccine expert, was the leader of this team.

Oral polio vaccines are now one of the greatest successes of modern science. They are administered throughout the world, and the World Health Organization predicts that early in the next century an extensive international vaccination campaign will have eradicated polio from all the world's populations. The vaccines used are those of Albert Sabin, who eventually won the race against Koprowski, but the two teams were very close for some time.

Back in the 1950s, Koprowski was impatient to test his vaccine on a large number of people, and since he had contacts with doctors in the Belgian Congo, he decided to launch a vaccine trial there. The Belgian authorities were obliging, and in those days this approval was all that was required. Over a period of three years, between 1957 and 1960, Koprowski and his colleagues fed his oral polio vaccine to approximately 1 million Africans in Congo, Rwanda, and Burundi.

Although this was a trial, the African subjects were not aware of it. All they knew was that their local chiefs had summoned them to line up before a group of white doctors who were dispensing what they were told were "sweets" but which were actually mouthfuls of vaccine squirted from a metal syringe. Koprowski and his Belgian colleagues kept a colony of a few hundred chimps at a research station in a remote part of the Congo, beside the Lindi River. They used these chimps in experiments to test the safety and effectiveness of the polio vaccines. However, Hooper believes that the vaccines used in Koprowski's African trials were contaminated with the primate precursor of HIV, and that it came from the chimpanzees at Lindi.

Oral polio vaccines are polio viruses that have been weakened so that they don't cause disease. In the fifties, scientists grew live polio vaccines in monkey kidneys that had been cut into tiny pieces and placed in a jar filled with fluid. Cells would grow out of the minced pieces of tissue and form a sort of underwater floor on the bottom of the dish. A few drops of a solution containing a weak vaccine strain of polio virus would be added to the dish; the virus would reproduce inside the cells, and after a week or so the fluid would be teeming with living polio vaccine virus. The fluid was filtered, diluted, and then fed to people. Their immune systems would learn to kill the vaccine virus, and when vaccinated people encountered the real virus later on, their bodies would be able to fight it off.

Hooper believes that Koprowski sent the kidneys of some chimpanzees from his Lindi research station to his lab in Philadelphia and used them to make the vaccine that was later administered in Africa. In *The River,* Hooper painstakingly tries to account for every chimp that was ever kept at Lindi in the late 1950s. He found that many died shortly after they were captured, and many others were killed after being used in experiments, but he claims that the fate of a few dozen chimps went unrecorded. If, as Hooper believes, the kidneys of these missing chimps were actually used to make Koprowski's vaccine, then it is possible that this vaccine might have contained the chimp version of HIV-1. The kidneys of all primates contain cells called macrophages, in which viruses like HIV can grow. Vaccine-making procedures in the 1950s were rather crude, so if Koprowski did use chimp kidneys, and if these kidneys came from a chimp carrying the precursor of HIV-1, he just might have been growing chimp HIV in his polio vaccine. When African children were fed the vaccine, the chimp virus might have infected some of them through oral cuts or through the mucous membranes in their mouths. When these children grew up, they would have passed the virus to their sexual partners and children, and before long, in Hooper's view, the global AIDS epidemic would be under way.

Versions of this story have been around for a while, although few scientists take it seriously anymore. It was first advanced in 1987 by a disqualified Texas medical doctor called Eva Lee Snead, who later wrote a book called *Some Call It "AIDS" — I Call It Murder! The Connection Between Cancer, AIDS, Immunizations and Genocide.* A philosopher named Louis Pascal heard Snead discussing AIDS and polio vaccines on the radio and decided to look into it further. He then wrote a detailed research paper arguing that it would have been technically possible for a primate precursor of HIV to have contaminated the Congo polio vaccines. He also criticized the scientific community and the editors of *The Lancet* and *Nature* for not publishing his articles. Later the journalist Tom Curtis wrote a more sober article about the polio theory in *Rolling Stone* magazine. Snead, Pascal, and Curtis argued that if AIDS had crept into the human race while hidden in a polio vaccine, it would not be the first time that something like this had happened. Many polio vaccines from the 1960s were contaminated with a monkey virus called SV40 that is now suspected of promoting various human cancers, including mesothelioma, an aggressive form of lung cancer.

Hooper has taken these arguments considerably further. Hooper is a journalist, not a scientist, but his 850-page book, with another 250 pages of notes, reflects some arduous reporting. He traveled back and forth across Europe and the United States, interviewing people ranging from famous AIDS researchers like David Ho and Robert Gallo, to the people who worked with Koprowski in the Congo, to a group of men sitting outside a bar in St. Louis who might have known someone who might have known someone who probably didn't die of AIDS in the 1960s. The detail is at times fascinating, at times confusing. Nevertheless, *The River* is among the best surveys to appear on the epidemiology of AIDS.

For Hooper's theory to be true, Koprowski must have been using chimpanzee kidneys to make the African vaccine. Koprowski claims that he grew his vaccine in the kidneys of Asian monkeys, not chimpanzees, and that his Congo chimps were used only for testing the vaccine. Everyone else who might have known which kidneys were used to make the vaccine is either dead or can't remember. Some people who were associated with the Congo research in the 1950s remember chimpanzee kidneys being shipped across the Atlantic to a lab across the street from Koprowski's in Philadelphia, although it is not clear that any of them went to Koprowski himself. Koprowski says his papers have been lost and there is no other documentation that might resolve the matter. Hooper found that the earliest cases of HIV and AIDS emerged from the very regions and, in some cases, the very towns and villages in Congo, Rwanda, and Burundi where some of Koprowski's trials took place. Could HIV have escaped from Koprowski's lab and spread throughout the world?*

Some scientists feel, like the sailor in *Island of Lost Souls,* that this is something we might not want to know. David Heymann of the World Health Organization told Curtis that "the origin of the AIDS virus is of no importance to scientists today." Another, rather irate AIDS researcher told him, "Who cares what the origin of the virus is? . . . It's distracting, it's non-productive, it's confusing to the public."

Is there any value in knowing where HIV came from? I think there is. HIV viruses crossed into human populations from primates at least twice, and perhaps more often, probably in the 1940s or 1950s, or perhaps a few decades earlier.† The only way to prevent other chimp or monkey viruses from doing so again is to know how such transfers could occur. And if a group of African scientists came to the United States and treated a million people with something that might, however improbably, have unleashed the most hideous plague in living memory, I think we'd want to know.

However, scientists who recognize that it is legitimate to ask how HIV came into being are skeptical of the theory that it came from a polio vaccine. John Moore, of the Aaron Diamond AIDS Research Center in New York, compared Hooper's polio-AIDS hypothesis to the grassy-knoll theory alleging that a second killer from the Mafia, the CIA, or some other agency, and not Lee Harvey Oswald, assassinated JFK. Koprowski himself is adamant that his vaccine could not possibly have started the AIDS epidemic. In October 1992 a com-

* HIV emerged at least twice in Africa, once around the Congo-Rwanda border as HIV-1 and once in West Africa as HIV-2. The two HIVs differ genetically, in that they have similar, but not identical, genes and those genes are ordered differently in the viral genome. Also, HIV-2 is less aggressive than HIV-1, in that it is transmitted less easily and people with HIV-2 develop AIDS more slowly than people with HIV-1. Koprowski's vaccine, according to Hooper, would only have been responsible for HIV-1. For the origin of HIV-2, Hooper postulates that Dr. Pierre Lepine of the Pasteur Institute conducted polio vaccine trials around Guinea-Bissau in the 1950s. Lepine claims to have grown his vaccine in baboon cells, but Hooper thinks that he might have used sooty mangabey and that this might have gone unrecorded. On the other hand, Hooper writes, Lepine might have used kidneys from a baboon that had been infected with a sooty mangabey virus.
† The first confirmed case of HIV infection was in 1959; it is possible that there were a few other cases in the decades before that went unrecorded, but it is unlikely that there were many.

mittee of AIDS experts investigated the polio vaccine hypothesis and exonerated Koprowski's vaccine. In December 1992 he sued *Rolling Stone* magazine and Tom Curtis for libel. The case was settled out of court.

The expert committee's decision had put the matter to rest for many observers, including me. But as I was reading *The River,* I recalled that about ten years ago another committee of experts in the United Kingdom concluded that beef from British cows was safe to eat, even though hundreds of thousands of cows were dropping dead from a terrifying and completely new brain disease called bovine spongiform encephalopathy (BSE). Forty-seven British people have already died from a new form of Creutzfeldt-Jakob disease, which they almost certainly got from eating meat from cows with BSE. Many more deaths are likely to follow over the coming decades. A few years before the BSE committee handed down its decision, another committee in France decided to delay the heat-treating of blood products for hemophiliacs and also the testing of blood for transfusions. Hundreds of people who might have escaped infection contracted HIV. So committees are hardly infallible, and sometimes, in order not to alarm or "confuse" the public, they put the public at risk.

2.

Many researchers believe there is a perfectly good explanation for where AIDS came from — that it was what Hooper calls "natural transfer." According to this theory, AIDS is really an old disease, and the HIV viruses were always fairly common in a small number of forest-dwelling communities that hunted monkeys and apes for food. Hunters subduing or butchering their prey would have become infected with the primate versions of HIV if they cut their hands and chimp or monkey blood seeped into the wounds. Because these tribes were isolated in remote villages, however, no one else ever became infected. Then upheavals in African society in the 1950s changed everything. After World War II the far-flung regions of the continent were suddenly drawn together as never before by highways, labor migrations, and refugee movements. It is plausible that these highways for people were also highways for germs.

The African wars of independence, the growth of African cities, new highways and truck routes, and the expansion of African mining industries suddenly drew men out of the countryside. Women, children, and the elderly were left behind in the villages, while their men looked for work, often hundreds of miles away, and prostitution flourished wherever these men went. Now the HIV viruses had many opportunities to escape from the bush, and every urban community, every truck stop and military barracks, was a breeding ground for HIV. All that was necessary for the virus to break out of the jungle was for a hunter or meat seller to cut his hand while butchering an infected chimp, and then for that hunter to migrate to a city or join an army and have sexual relations, perhaps in a brothel. The virus might then spread from the prostitute who had sex with the hunter to other customers, and then perhaps to other brothels visited by those customers, and then to yet more customers, and eventually to the wives of customers. Most HIV-infected people do not get sick for five or ten years after they are infected. During that time they can unknowingly spread the virus to others.

This is a plausible theory, even though there is something rather Victorian about it. In the nineteenth century, theories about the impact of modern urban life on fragile souls from traditional tribal villages were sometimes invoked to explain why Africans were so susceptible to tuberculosis. Perhaps, for some people, the natural transfer theory suggests that AIDS is the price Africans have paid for independence, war, urban drift, sexual license, and being cruel to chimps and monkeys.

Hooper is skeptical about the natural transfer hypothesis for other reasons, and his arguments are worth considering. Indeed, I found *The River* most interesting where Hooper examines, and rejects, the natural transfer theory, which has been widely accepted since the 1980s. For one thing, Hooper argues, the natural transfer theory does not really explain why AIDS emerged when it did. Africans have been killing and eating monkeys for at least 50,000 years, and yet African and colonial doctors, even those who worked in rural health centers, had never seen anything like AIDS before. Moreover, there is no reason to believe that colonial or precolonial Africa was all quiet villages and stable families. Certainly prostitution existed in the nineteenth century, as did extensive trade between the interior and the coasts. Since the sixteenth century, Africans have engaged increasingly in war, sometimes with European invaders and sometimes among themselves. For example, in the early nineteenth century, land disputes among rival chiefs in southern Africa ignited the "Wars of Wandering," named for the extensive migrations that followed.

The slave trade, which reached deep into the interior, existed in Africa long before colonial times and accelerated after the arrival of the Portuguese in the fifteenth century. It has been estimated that between 1700 and 1850 alone, 21 million Africans were enslaved, and at least 9 million were marched to the coasts and shipped all over the world. During the two World Wars, African men were recruited as soldiers to fight in North Africa and the Middle East. These migrations did spread HTLV-1, a virus similar to HIV that also seems to have come from chimpanzees, as well as malaria and yellow fever. So why not HIV?

Hooper also questions the natural transfer theory because he believes that more than a hunter's wound must have started the AIDS epidemic. HIV-1 is largely absent from the Pygmy communities that still live in forests and hunt the very chimpanzees that carry viruses closely related to HIV-1. Pygmy hunters use rough tools to butcher their prey, so if the natural transfer theory is correct, Pygmies, if anyone, should be infected with HIV. However, the only HIV-positive Pygmies identified so far are those who have had significant contacts with larger towns and probably picked up the virus through sexual intercourse. Pygmies do seem to carry HTLV-1, which also spreads through blood and comes from chimpanzees. If Pygmies can contract HTLV-1 from chimps, why don't they contract HIV?

One reason may be that the monkey and chimp versions of HIV may not be harmful to human beings, and may not spread from one person to another. In 1990 a laboratory worker became infected with a monkey virus related to the one that is thought to have given rise to HIV-2 in West Africa. He (or she — the sex of the lab worker was not reported) was working with the blood of a macaque monkey that had been infected with an HIV-like virus from a sooty mangabey. It is not clear exactly how the lab worker became infected, but he had been suffering from a case of poison ivy and the rubber lab gloves hurt his hands, so he didn't wear them. It is likely that he spilled something — blood or some other fluid with virus in it — on his hands and that the virus seeped in through the sores. The lab worker's immune system made antibodies against the virus, but the virus itself seemed to grow very slowly. The lab worker is still healthy, almost ten years after being infected. Similarly, several Africans, including two Liberian rubber plantation workers and a woman who sold monkey meat at a market in Sierra Leone, were infected with the sooty mangabey virus thought to have given rise to HIV-2; but again, the infections seem not to have progressed. These people never got sick, and their viruses seem not to have spread to anyone else. This is virtually unheard of with real HIV infection, which never clears and, as far as is known, always ends in death.

Something probably did happen in central Africa before the 1960s, and perhaps simultaneously in West Africa, to cause a very small number of monkey viruses that were previously harmless to human beings to become deadly, to change their "chemical rhythm," as Dr. Moreau would say. Somehow they evolved the ability to creep into semen and other secretions, and spread from person to person, and somehow they began to grow so rapidly in human blood that they were able to overwhelm the immune system and destroy it.

What happened, exactly, is still a mystery, and Hooper's polio vaccine theory may not explain it either. First of all, there is no proof, or even a very strong indication, that chimp kidneys were used to make the vaccine administered in Africa. The only person who remembers that chimp kidneys even went to Koprowski's lab is the wife of his technician, who admits that her memory of the period is hazy. She does tell Hooper she is pretty sure some chimp kidneys went to Koprowski's lab, but since no one else remembers this, and since there is no record of what Koprowski did with the kidneys, this is still not much to go on. Moreover, yet another deadly HIV virus may have emerged since the polio vaccine trials ended. The HIV variant known as HIV-1N, which is extremely rare and seems to be found only in Cameroon, may have been transferred from chimps to people only recently.

Hooper's theory also does not account for the crucial step, the mutation of the monkey virus into the aggressive, deadly human virus, HIV. If butchers of monkey meat, Pygmies who hunt chimpanzees, and lab technicians who spill monkey blood on their hands

seem not to get AIDS, why would people get it from monkey viruses
in a polio vaccine?

3.

What did happen? How did an apparently harmless monkey virus
turn into a killer? While writing this article, I asked several AIDS researchers where they thought HIV came from. Just about all of
them said that they thought it was an old African disease that had
emerged from the bush when the winds of change blew through
the continent — that HIV was driven by urbanization, war, truck-
ing, mining, and prostitution. While I agreed the virus might be
spreading that way, I told them I also agreed with Hooper's view
that AIDS just didn't look like an old African disease. It looked like
HIV had crossed from primates to people quite recently.

One hypothesis, which Hooper pursues only briefly in *The River,*
intrigued me. A very small number of scientists have quietly been
suggesting that HIV may have first started spreading (and may still
be spreading) into human populations through unsterilized needles used in doctors' clinics or in vaccination campaigns, or even in
blood transfusions.* While I was talking to Patricia Fultz, who studies HIV-like viruses in monkeys at the University of Alabama, she
mentioned the work of Opendra Narayan at the University of Kansas. I discovered that Narayan had actually succeeded in turning an
apparently harmless monkey virus into an AIDS virus in his own
laboratory.

Narayan works with a genetically engineered virus called a SHIV
(simian human immunodeficiency virus), which is a version of HIV
that is used to infect lab monkeys. This SHIV grows in monkeys but
does not cause disease in those monkeys, and it is not passed to
other monkeys through sex, biting, or other natural means. However, this SHIV can be passed from one monkey to another artificially in laboratories, for example through blood transfusions or

---

* See Zhu et al., "An African HIV-1 Sequence from 1959 and Implications for the
Origin of the Epidemic," *Nature,* Vol. 391 (1998), pp. 594–597. This article contains the following aside: "The factors that propelled the initial spread of HIV-1 in
central Africa remain unknown: the role of large-scale vaccination campaigns, perhaps with multiple uses of non-sterilized needles, should be carefully examined, although social changes such as easier access to transportation, increasing population
density and more frequent sexual contacts may have been more important."

through bone marrow transplantation. If the SHIV is transmitted artificially from one monkey to another rapidly enough, through a process known as "passaging," it can turn into a virus that spreads easily and causes AIDS in monkeys.

"This has been known since Pasteur's time," Narayan told me. "If you take any virus and 'passage' it through a new species often enough, eventually you get a more pathogenic virus." To prove that this would work with SHIV, Narayan and his colleagues injected it into a monkey and waited for it to grow. Usually the monkey's immune system controls the virus and clears it, but this takes a month or so. After a few weeks, Narayan took the virus from the monkey's bone marrow, where it was still growing, and then injected it into another monkey. A few weeks later he took the virus from the second monkey's bone marrow and injected it into a third monkey. The virus caused no disease in the first two monkeys, but it caused mild disease in the third monkey. But when, after a few more weeks, the third monkey's virus was passed to a fourth monkey, the virus caused AIDS in almost every subsequent monkey Narayan injected with it.

Something similar may have happened with HIV. If a hunter or monkey-meat butcher became infected with a harmless monkey virus and then shortly afterward passed it on to someone else, who then passed it on to someone else a few weeks later, it is possible that the monkey virus might have turned into HIV. Like Narayan's SHIV, the monkey virus might not have been able to cross from the hunter to other people by sex, but it just might have been able to cross to others through blood. This might have happened in the clinics and hospitals of twentieth-century Africa.

Perhaps a hunter or butcher carrying a benign monkey virus gave blood at a blood bank or had an injection. Perhaps someone was transfused with his blood, or perhaps the needle used to inject him was used to inject someone else without being sterilized. Perhaps, a few weeks later, the virus was transferred to a third person through another injection or transfusion. This might have been enough to "kick-start" the virus. It might have evolved through such passaging to become able to grow vigorously in human cells. It might have been able to infect new people through means other than needles or blood transfusions. It might have become sexually

transmitted, and it might have become deadly. Hypodermic needles were introduced in Africa in the early part of this century,
and blood banks were introduced later, probably after World
War II. I asked Dr. Narayan whether primate HIVs might have first
adapted to human beings after being passaged through blood
transfusions or unsterilized needles, just as his SHIV adapted to
monkeys through successive bone marrow transfers. "Yes," he said,
"that might have happened."

Such passaging events might be very rare. "We used to talk in
terms of lightning rods," the AIDS researcher Preston Marx told
Hooper. "You know — lightning can't strike twice unless there's a
lightning rod. The lightning rod's the needle. That's why it struck
twice in the same place — HIV-1 in Central Africa and HIV-2 in
West Africa." Hooper dismisses the hypothesis that hypodermic
needles or blood transfusions had anything to do with the origin of
HIV, but he does not consider the work of Dr. Narayan and his colleagues.

In our efforts to make things better for ourselves, we sometimes
make things worse. If HIV entered human populations through
such new medical technologies as needles, blood transfusions, or
even polio vaccines, it will not be the first time that a microbe has
flourished in the wake of scientific advance. Smallpox, brucellosis,
anthrax, and tuberculosis are all cattle diseases, influenza comes
from hogs, leprosy from the water buffalo, the common cold from
horses, and measles, rabies, and hydatid cysts from dogs. All these
diseases probably first affected human beings after these animals
were first domesticated and when the density of human populations increased sufficiently to permit these microbes to propa-
gate. Epidemics of polio only emerged in the nineteenth century
with improvements in sanitation. Animal husbandry and sanitation
have saved generations from malnutrition, dysentery, cholera, and
many other plagues, only for new ones to emerge in their places.

4.

Now that the AIDS epidemic is under way, a vital question that remains is why the disease is now overwhelming so many developing
countries and remains a serious problem in the United States, es-

pecially among the poor. In 1998, 5.8 million people contracted HIV, more than two thirds of them in sub-Saharan Africa. Asia is catching up: 7 million Indians now carry HIV, as may as many as 1 million Chinese. In the United States, death rates from AIDS are falling as a result of the success of new medications, but the number of new infections seems stable — and may even be rising, especially among women and people of color.

Many governments have been ambivalent about addressing the AIDS epidemic. Countries such as Uganda and Thailand that have invested in comprehensive prevention and information campaigns have saved thousands of lives; still, the virus persists, even in communities that seem well informed. A 1995 survey from Uganda showed that more than two thirds of those interviewed knew how to prevent transmission of HIV, for example through the use of condoms. A simultaneous survey showed that fewer than 5 percent of women had ever used a condom in their lives. Condoms are expensive for people in developing countries, sometimes poorly made, and often hard to find. Men don't like them, and it is often impossible for women to insist they be used.

Manuel Carballo, who helped pioneer the World Health Organization's AIDS program in the 1980s, told me he believes HIV "follows the contours of inequality." Early in the epidemic, there was a high incidence of AIDS among relatively prosperous people, including professional people in Africa, and among gay men in the United States. But AIDS is becoming increasingly a disease of the poor, both in the developing world and in the West. Poverty, powerlessness, and exclusion are perhaps the greatest risk factors for HIV infection today.

In Uganda a few years ago, I went to see a rehearsal of a rather dark musical comedy put on by a local AIDS organization. The central character and his wife lived in a village near Kampala with their four children. They were poor, but he wanted his wife to have another baby. She refused, and wanted him to use condoms. He went to the city, where he met two girls in a bar. They were even poorer than he was and behind his back they sang about how they looked forward to filling their bellies. One of them became pregnant, and then the man brought both girls back to his house in the village. The girls gave the man HIV, which he transmitted to his wife. In the end, the man, his wife, the two girls, and the girl's baby all died, and the man's relatives all came together in a rousing finale to fight over his property.

The audience consisted of a few locals, friends of the actors, and others who just stopped by to watch. An American woman who worked for the AIDS organization wanted to know what people had thought of the play. "What did you think the message was?" she asked, addressing the audience. "Don't go out with bar girls," someone said. "Stick to one partner," said another. Then an older woman spoke up. She worked as a healer and herbalist, and she was wearing a voluminous green basuti, the traditional costume of her tribe. "AIDS has come to haunt a world that thought it was incomplete," she said. "Some wanted children, some wanted money, some wanted property, and all we ended up with is AIDS."

ANNE FADIMAN

Under Water

FROM *The New Yorker*

WHEN I WAS EIGHTEEN, I was a student on a month-long wilderness program in western Wyoming. On the third day, we went canoeing on the Green River, a tributary of the Colorado that begins in the glaciers of the Wind River Range and flows south across the sagebrush plains. Swollen by warm-weather runoff from an unusually deep snowpack, the Green was higher and swifter that month — June of 1972 — than it had been in forty years. A river at flood stage can have strange currents. There is not enough room in the channel for the water to move downstream in an orderly way, so it collides with itself and forms whirlpools and boils and souse holes. Our instructors decided to stick to their itinerary nevertheless, but they put in at a relatively easy section of the Green, one that the flood had merely upgraded, in the international system of whitewater classification, from Class I to Class II. There are six levels of difficulty, and Class II was not an unreasonable challenge for novice paddlers.

The Green River did not seem dangerous to me. It seemed magnificently unobstructed. Impediments to progress — the rocks and stranded trees that under normal conditions would protrude above the surface — were mostly submerged. The river carried our aluminum canoe high and lightly, like a child on a broad pair of shoulders. We could rest our paddles on the gunwales and let the water do our work. The sun was bright and hot. Every few minutes, I dipped my bandanna in the river, draped it over my head, and let an ounce or two of melted glacier run down my neck.

I was in the bow of the third canoe. We rounded a bend and saw, fifty feet ahead, a standing wave in the wake of a large black boulder. The students in the lead canoe were backferrying, slipping crabwise across the current by angling their boat diagonally and stroking backward. Backferrying allows paddlers to hover midstream and carefully plan their course instead of surrendering to the water's pace. But if they lean upstream — a natural inclination, for few people choose to lean toward the difficulties that lie ahead — the current can overflow the lowered gunwale and flip the boat. And that is what happened to the lead canoe.

I wasn't worried when I saw it go over. Knowing that we might capsize in the fast water, our instructors had arranged to have our gear trucked to our next campsite. The packs were all safe. The water was little more than waist-deep, and the paddlers were both wearing life jackets. They would be fine. One was already scrambling onto the right-hand bank.

But where was the second paddler? Gary, a local boy from Rawlins, a year or two younger than I, seemed to be hung up on something. He was standing at a strange angle in the middle of the river, just downstream from the boulder. Gary was the only student on the course who had not brought sneakers, and one of his mountaineering boots had become wedged between two rocks. The other canoes would come around the bend in a moment, and the instructors would pluck him out.

But they didn't come. The second canoe pulled over to the bank and ours followed. Thirty seconds passed, maybe a minute. Then we saw the standing wave bend Gary's body forward at the waist, push his face underwater, stretch his arms in front of him, and slip his orange life jacket off his shoulders. The life jacket lingered for a moment at his wrists before it floated downstream, its long white straps twisting in the current. His shirtless torso was pale and undulating, and it changed shape as hills and valleys of water flowed over him, altering the curve of the liquid lens through which we watched him. I thought, He looks like the flayed skin of St. Bartholomew in the Sistine Chapel. As soon as I had the thought, I knew that it was dishonorable. To think about anything outside the moment, outside Gary, was a crime of inattention. I swallowed a small, sour piece of self-knowledge: I was the sort of person who, instead of weeping or shouting or praying during a crisis, thought about something from a textbook (H. W. Janson's *History of Art,* page 360).

Once the flayed man had come, I could not stop the stream of images: Gary looked like a piece of seaweed, Gary looked like a waving handkerchief, Gary looked like a hula dancer. Each simile was a way to avoid thinking about what Gary was, a drowning boy. To remember these things is dishonorable, too, for I have long since forgotten Gary's last name and the color of his hair and the sound of his voice.

I do not remember a single word that anyone said. Somehow, we got into one of the canoes, all five of us, and tried to ferry the twenty feet or so to the middle of the river. The current was so strong, and we were so incompetent, that we never got close. Then we tried it on foot, linking arms to form a chain. The water was so cold that it stung. And it was noisy — not the roar and crash of white water but a groan, a terrible bass grumble, from the stones that were rolling and leaping down the riverbed. When we got close to Gary, we couldn't see him; all we could see was the reflection of the sky. A couple of times, groping blindly, one of us touched him, but he was as slippery as soap. Then our knees buckled and our elbows unlocked, and we rolled downstream, like the stones. The river's rocky load, moving invisibly beneath its smooth surface, pounded and scraped us. Eventually, the current heaved us, blue-lipped and panting, onto the bank. In that other world above the water, the only sounds were the buzzing of bees and flies. Our wet sneakers kicked up red dust. The air smelled of sage and rabbitbrush and sunbaked earth.

We tried again and again, back and forth between the worlds. Wet, dry, cold, hot, turbulent, still.

At first, I assumed that we would save him. He would lie on the bank and the sun would warm him while we administered mouth-to-mouth resuscitation. If we couldn't get him out, we would hold him upright in the river, and maybe he could still breathe. But the Green River was flowing at nearly three thousand cubic feet — about ninety tons — per second. At that rate, water can wrap a canoe around a boulder like tinfoil. Water can uproot a tree. Water can squeeze the air out of a boy's lungs, undo knots, drag off a life jacket, lever a boot so tightly into the riverbed that even if we had had ropes — the ropes that were in the packs that were in the trucks — we could never have budged him.

We kept going in, not because we had any hope of rescuing Gary after the first ten minutes, but because we had to save face. It would have been humiliating if the instructors came around the bend and found us sitting in the sagebrush, a docile row of five with no hypothermia and no skinned knees. Eventually, they did come. The boats had been delayed because one had nearly capsized, and the instructors had made the other students stop and practice backferrying until they learned not to lean upstream. Even though Gary had already drowned, the instructors did all the same things we had done, more competently but no more effectively, because they, too, would have been humiliated if they hadn't skinned their knees. Men in wet suits, belayed with ropes, pried the body out the next morning.

When I was eighteen, I wanted to hurry through life as fast as I could. Twenty-seven years have passed, and my life now seems too fast. I find myself wanting to backferry, to hover midstream, suspended. I might then avoid many things: harsh words, foolish decisions, moments of inattention, regrets that wash over me, like water.

The Cancer-Cluster Myth

FROM *The New Yorker*

Is it something in the water? During the past two decades, reports of cancer clusters — communities in which there seems to be an unusual number of cancers — have soared. The place-names and the suspects vary, but the basic story is nearly always the same. The Central Valley farming town of McFarland, California, came to national attention in the eighties after a woman whose child was found to have cancer learned of four other children with cancer in just a few blocks around her home. Soon doctors identified six more cases in the town, which had a population of 6,400. The childhood-cancer rate proved to be four times as high as expected. Suspicion fell on groundwater wells that had been contaminated by pesticides, and lawsuits were filed against six chemical companies.

In 1990, in Los Alamos, New Mexico, a local artist learned of seven cases of brain cancer among residents of a small section of the town's Western Area. How could seven cases of brain cancer in one neighborhood be merely a coincidence? "I think there is something seriously wrong with the Western Area," the artist, Tyler Mercier, told the *Times*. "The neighborhood may be contaminated." In fact, the Los Alamos National Laboratory, which was the birthplace of the atomic bomb, had once dumped millions of gallons of radioactive and toxic waste in the surrounding desert, without providing any solid documentation about precisely what was dumped or where. In San Ramon, California, a cluster of brain cancers was discovered at a high-school class reunion. On Long Island, federal, state, and local officials are currently spending $21

million to try to find out why towns like West Islip and Levittown
have elevated rates of breast cancer.

I myself live in a cancer cluster. A resident in my town — New-
ton, Massachusetts — became suspicious of a decades-old dump
next to an elementary school after her son developed cancer. She
went from door to door and turned up forty-two cases of cancer
within a few blocks of her home. The cluster is being investigated
by the state health department.

No doubt, one reason for the veritable cluster of cancer clusters
in recent years is the widespread attention that cases like those in
McFarland and Los Alamos received, and the ensuing increase in
public awareness and concern. Another reason, though, is the way
in which states have responded to that concern: they've made available to the public data on potential toxic sites, along with information from "cancer registries" about local cancer rates. The result
has been to make it easier for people to find worrisome patterns,
and, more and more, they've done so. In the late eighties, public-
health departments were receiving between 1,300 and 1,600 reports of feared cancer clusters, or "cluster alarms," each year. Last
year, in Massachusetts alone, the state health department responded to between 3,000 and 4,000 cluster alarms. Under public
pressure, state and federal agencies throughout the country are engaging in "cancer mapping" to find clusters that nobody has yet reported.

A community that is afflicted with an unusual number of cancers quite naturally looks for a cause in the environment — in the
ground, the water, the air. And correlations are sometimes found:
the cluster may arise after, say, contamination of the water supply
by a possible carcinogen. The problem is that when scientists have
tried to confirm such causes, they haven't been able to. Raymond
Richard Neutra, California's chief environmental health investigator and an expert on cancer clusters, points out that among hundreds of exhaustive, published investigations of residential clusters
in the United States, not one has convincingly identified an underlying environmental cause. Abroad, in only a handful of cases has a
neighborhood cancer cluster been shown to arise from an environmental cause. And only one of these cases ended with the discovery
of an unrecognized carcinogen. It was in a Turkish village called
Karain, where twenty-five cases of mesothelioma, a rare form of

lung cancer, cropped up among fewer than eight hundred villagers. (Scientists traced the cancer to a mineral called erionite, which is abundant in the soil there.) Given the exceedingly poor success rate of such investigations, epidemiologists tend to be skeptical about their worth.

When public-health investigators fail to turn up any explanation for the appearance of a cancer cluster, communities can find it frustrating, even suspicious. After all, these investigators are highly efficient in tracking down the causes of other kinds of disease clusters. "Outbreak" stories usually start the same way: someone has an intuition that there are just too many people coming down with some illness and asks the health department to investigate. With outbreaks, though, such intuitions are vindicated in case after case. Consider the cluster of American Legionnaires who came down with an unusual lung disease in Philadelphia in 1976; the startling number of limb deformities among children born to Japanese women in the sixties; and the appearance of rare *Pneumocystis carinii* pneumonia in five young homosexual men in Los Angeles in 1981. All these clusters prompted what are called "hot-pursuit investigations" by public-health authorities, and all resulted in the definitive identification of a cause: namely, *Legionella* pneumonitis, or Legionnaires' disease; mercury poisoning from contaminated fish; and HIV infection. In fact, successful hot-pursuit investigations of disease clusters take place almost every day. A typical recent issue of the Centers for Disease Control's *Morbidity and Mortality Weekly Report* described a cluster of six patients who developed muscle pain after eating fried fish. Investigation by health authorities identified the condition as Haff disease, which is caused by a toxin sometimes present in buffalo fish. Four of the cases were traced to a single Louisiana wholesaler, whose suppliers fished the same tributaries of the Mississippi River.

What's more, for centuries scientists have succeeded in tracking down the causes of clusters of cancers that aren't residential. In 1775 the surgeon Percivall Pott discovered a cluster of scrotal-cancer cases among London chimney sweeps. It was common practice then for young boys to do their job naked, the better to slither down chimneys, and so high concentrations of carcinogenic coal dust would accumulate in the ridges of their scrota. Pott's chimney sweeps proved to be a classic example of an "occupational" cluster. Scientists have also been successful in investigating so-called medical clusters. In the late 1960s, for example, the pathologist Arthur Herbst was surprised to come across eight women between the ages of fifteen and twenty-two who had clear-cell adenocarcinoma, a type of cervical cancer that had never been seen in women so young. In 1971 he published a study linking the cases to an anti-miscarriage drug called diethylstilbestrol, or DES, which the mothers of these women had taken during pregnancy. Subsequent studies confirmed the link with DES, which was taken by some 5 million pregnant women between 1938 and 1971. The investigation of medical and occupational cancer clusters has led to the discovery of dozens of carcinogens, including asbestos, vinyl chloride, and certain artificial dyes.

So why don't hot-pursuit investigations of neighborhood cancer clusters yield such successes? For one thing, many clusters fall apart simply because they violate basic rules of cancer behavior. Cancer develops when a cell starts multiplying out of control, and the process by which this happens isn't straightforward. A carcinogen doesn't just flip some cancer switch to "on." Cells have a variety of genes that keep them functioning normally, and it takes an almost chance combination of successive mutations in these genes — multiple "hits," as cancer biologists put it — to make a cell cancerous rather than simply killing it. A carcinogen provides one hit. Other hits may come from a genetic defect, a further environmental exposure, a spontaneous mutation. Even when people have been subjected to a heavy dose of a carcinogen and many cells have been damaged, they will not all get cancer. (For example, DES causes clear-cell adenocarcinoma in only one out of a thousand women exposed to it in utero.) As a rule, it takes a long time before a cell receives enough hits to produce the cancer, and so, unlike infections or acute toxic reactions, the effect of a carcinogen in a community won't be seen for years. Besides, in a mobile society like ours, cancer victims who seem to be clustered may not all have lived in an area long enough for their cancers to have a common cause.

To produce a cancer cluster, a carcinogen has to hit a great many cells in a great many people. A brief, low-level exposure to a carcinogen is unlikely to do the job. Raymond Richard Neutra has calculated that for a carcinogen to produce a sevenfold increase in the occurrence of a cancer (a rate of increase not considered particularly high by epidemiologists) a population would have to be exposed to 70 percent of the maximum tolerated dose in the course of a full year, or the equivalent. "This kind of exposure is credible as part of chemotherapy or in some work settings," he wrote in a 1990 paper, "but it must be very rare for most neighborhood and school settings." For that reason, investigations of occupational cancer clusters have been vastly more successful than investigations of residential cancer clusters.

Matters are further complicated by the fact that cancer isn't one disease. What turns a breast cell into breast cancer isn't what turns a white blood cell into leukemia: the precise combination of hits varies. Yet some clusters lump together people with tumors that have entirely different biologies and are unlikely to have the same cause. The cluster in McFarland, for example, involved eleven children with nine kinds of cancer. Some of the brain-cancer cases in the Los Alamos cluster were really cancers of other organs that had metastasized to the brain.

If true neighborhood clusters — that is, local clusters arising from a common environmental cause — are so rare, why do we see so many? In a sense, we're programmed to: nearly all of them are the result of almost irresistible errors in perception. In a pioneering article published in 1971, the cognitive psychologists Daniel Kahneman and Amos Tversky identified a systematic error in human judgment, which they called the Belief in the Law of Small Numbers. People assume that the pattern of a large population will be replicated in all its subsets. But clusters will occur simply through chance. After seeing a long sequence of red on the roulette wheel, people find it hard to resist the idea that black is "due" — or else they start to wonder whether the wheel is rigged. We assume that a sequence of R-R-R-R-R-R is somehow less random than, say, R-R-B-R-B-B. But the two sequences are equally likely. (Casinos make a lot of money from the Belief in the Law of Small Numbers.) Truly random patterns often don't appear random to us. The statistician William Feller studied one classic example. During the Germans' intensive bombing of South London in the Second World War, a few areas were hit several times and others were not hit at all.

The places that were not hit seemed to have been deliberately spared, and, Kahneman says, people became convinced that those places were where the Germans had their spies. When Feller analyzed the statistics of the bomb hits, however, he found that the distribution matched a random pattern.

Daniel Kahneman himself was involved in a similar case. "During the Yom Kippur War, in 1973, I was approached by people in the Israeli Air Force," he told me. "They had two squads that had left base, and when the squads came back one had lost four planes and the other had lost none. They wanted to investigate for all kinds of differences between the squadrons, like whether pilots in one squadron had seen their wives more than in the other. I told them to stop wasting their time." A difference of four lost planes could easily have occurred by chance. Yet Kahneman knew that if Air Force officials investigated they would inevitably find some measurable differences between the squadrons and feel compelled to act on them.

Human beings evidently have a deep-seated tendency to see meaning in the ordinary variations that are bound to appear in small samples. For example, most basketball players and fans believe that players have hot and cold streaks in shooting. In a paper entitled "The Hot Hand in Basketball," Tversky and two colleagues painstakingly analyzed the shooting of individual players in more than eighty games played by the Philadelphia 76ers, the New Jersey Nets, and the New York Knicks during the 1980–1981 season. It turned out that basketball players — even notorious "streak shooters" — have no more runs of hits or misses than would be expected by chance. Because of the human tendency to perceive clusters in random sequences, however, Tversky and his colleagues found that "no amount of exposure to such sequences will convince the player, the coach, or the fan that the sequences are in fact random. The more basketball one watches and plays, the more opportunities one has to observe what appears to be streak shooting."

In epidemiology, the tendency to isolate clusters from their context is known as the Texas sharpshooter fallacy. Like a Texas sharpshooter who shoots at the side of a barn and then draws a bull's-eye around the bullet holes, we tend to notice cases first — four cancer patients on one street — and then define the population base around them. With rare conditions, such as Haff disease or mercury poisoning, even a small clutch of cases really would represent a dramatic excess, no matter how much Texas sharp-shooting we did. But most cancers are common enough that noticeable residential clusters are bound to occur. Raymond Richard Neutra points out that given a typical registry of eighty different cancers, you could expect 2,750 of California's 5,000 census tracts to have statistically significant but perfectly random elevations of cancer. So if you check to see whether your neighborhood has an elevated rate of a specific cancer, chances are better than even that it does — and it almost certainly won't mean a thing. Even when you've established a correlation between a specific cancer and a potential carcinogen, scientists have hardly any way to distinguish the "true" cancer cluster that's worth investigating from the crowd of cluster impostors.

One helpful tip-off is an extraordinarily high cancer rate. In Karain, Turkey, the incidence of mesothelioma was more than *seven thousand times* as high as expected. In even the most serious cluster alarms that public-health departments have received, however, the cancer rate has been nowhere near that high. (The lawyer Jan Schlichtmann, of *Civil Action* fame, is now representing victims of a cancer cluster in Dover Township, New Jersey, where the child-hood-cancer rate is 30 percent higher than expected.)

This isn't to say that carcinogens in the local environment can't raise cancer rates; it's just that such increases disappear in all the background variation that occurs in small populations. In larger populations, it's a different story. The 1986 Chernobyl disaster exposed hundreds of thousands of people to radiation; scientists were able to establish that it caused a more than one-hundredfold increase in thyroid cancer among children years later. By contrast, investigating an isolated neighborhood cancer cluster is almost always a futile exercise. Investigators knock on doors, track down former residents, and check medical records. They sample air, soil, and water. Thousands, sometimes millions, of dollars are spent. And with all those tests, correlations inevitably turn up. Yet, years later, in case after case, nothing definite is confirmed.

"The reality is that they're an absolute, total, and complete waste of taxpayer dollars," says Alan Bender, an epidemiologist with the Minnesota Department of Health, which investigated more than

1,000 cancer clusters in the state between 1984 and 1995. The problem of perception and politics, however, remains. If you're a public-health official, try explaining why a dozen children with cancer in one neighborhood doesn't warrant investigation. According to a national study, health departments have been able to reassure people by education in more than 70 percent of cluster alarms. Somewhere between 1 and 3 percent of alarms, however, result in expensive on-site investigations. And the cases that are investigated aren't even the best-grounded ones: they are the cases pushed by the media, enraged citizens, or politicians. "Look, you can't just kiss people off," Bender says. In fact, Minnesota has built such an effective public-response apparatus that it has not needed to conduct a formal cluster investigation in three years.

Public-health departments aren't lavishly funded, and scientists are reluctant to see money spent on something that has proved to be as unproductive as neighborhood cluster alarms or cancer mapping. Still, public confidence is poorly served by officials who respond to inquiries with a scientific brushoff and a layer of bureaucracy. To be part of a cancer cluster is a frightening thing, and it magnifies our ordinary response when cancer strikes: we want to hold something or someone responsible, even allocate blame. Health officials who understand the fear and anger can have impressive success, as the ones in Minnesota have shown. But there are times when you cannot maintain public trust without acting on public concerns. Science alone won't put to rest questions like the one a McFarland mother posed to the *Los Angeles Times:* "How many more of our children must die before something is done?"

BRIAN HAYES

Clock of Ages

FROM *The Sciences*

As THE WORLD spirals on toward 01-01-00, survivalists are hoarding cash, canned goods, and shotgun shells. It's not the Rapture or the Revolution they await, but a technological apocalypse. Y2K! The lights are going out, they warn. Banks may fail. Airplanes may crash. Your VCR will go on the blink. Who could have foreseen such turmoil? Decades back, one might have predicted anxiety and unrest at the end of the millennium, but no one could have guessed that the cause would be an obscure shortcut written into computer software by unknown programmers in the 1960s and 1970s.

Whether or not civilization collapses on January 1, those programmers do seem, in hindsight, to have been pretty short on foresight. How could they have failed to look beyond year '99? But I give them the benefit of the doubt. All the evidence suggests they were neither stupid nor malicious. What led to the Y2K bug was not arrogant indifference to the future ("I'll be retired by then. Let the next shift fix it"). On the contrary, it was an excess of modesty ("No way *my* code will still be running thirty years out"). The programmers could not envision that their hurried hacks and kludges would become the next generation's "legacy systems."

Against this background of throwaway products that someone forgot to throw away, it may be instructive to reflect on a computational device built in a much different spirit. This machine was carefully crafted for Y2K compliance, even though it was manufactured when the millennium was still a couple of lifetimes away. As a matter of fact, the computer is equipped to run until the year 9999

— and perhaps even beyond, with a simple Y10K patch. This achievement might serve as an object lesson to the software engineers of the present era. But I am not quite sure just what the lesson is.

The machine I speak of is the astronomical clock of Strasbourg Cathedral, built and rebuilt several times in the past 600 years. The present version is a nineteenth-century construction, still ticking along smartly at age 150-something. If all goes as planned, it will navigate the various calendrical cataracts of the coming months without incident, unfazed by January 1, 2000, or the subsequent February 29, or the revels of the latter-day millenarians on New Year's Day, 2001.

The Strasbourg Cathedral clock is not a tower clock, like Big Ben in London, meant to broadcast the hours to the city. It stands inside the cathedral in a case of carved stone and wood 50 feet high and 24 feet wide, with three ornamented spires and a gigantic instrument panel of dials and globes, plus a large cast of performing automata. Inside the clock is a glory of gears.

"Clock" is hardly an adequate description. More than a timepiece, it is an astronomical and calendrical computer. A celestial globe in front of the main cabinet tracks the positions of 5,000 stars, while a device much like an orrery models the motions of the six innermost planets. The current phase of the moon is indicated by a rotating globe, half gilt and half black.

If you want to know what time it is, the clock offers a choice of answers. A dial mounted on the celestial globe shows sidereal time, as measured by the earth's rotation with respect to the fixed stars. A larger dial on the front of the clock indicates local solar time, which is essentially what a sundial provides; the prick of noon by that measure always comes when the sun is highest overhead. The pointer for local lunar time is similarly synchronized to the height of the moon. Still another dial, with familiar-looking hour and minute hands, shows mean solar time, which averages out the seasonal variations in the earth's orbital velocity to make all days equal in length, exactly twenty-four hours. A second pair of hands on the same dial shows civil time, which in Strasbourg runs thirty minutes ahead of mean solar time.

There's more. A golden wheel nine feet in diameter, marked off into 365 divisions, turns once a year, while Apollo stands at one side to point out today's date. What about leap years? Presto: an extra day magically appears when needed. Each daily slot on the calendar wheel is marked with the name of a saint or some church occasion. Of particular importance is the inclusion of Easter and the other "movable feasts" of the ecclesiastical calendar. Calculating the dates of those holidays requires feats of mechanical trickery.

For the Y2K police, the crucial component of the clock is, of course, the counter of years. It is an inconspicuous four-digit register that anyone from our age of automobiles will instantly recognize as an odometer. On December 31, at midnight, mean solar time — and thus half an hour late by French official time — the digits will roll over from one-triple-nine to two-triple-zero.

Wait! There's even more! The clock is inhabited by enough animated figures to open a small theme park. The day of the week is marked by a slow procession of seven Greco-Roman gods in chariots. Each day at noon (that's mean solar noon), the twelve apostles appear, saluting a figure of Christ, who blesses each in turn. Every hour a putto overturns a sandglass. At various other times figures representing the four ages of man and a skeletal Death emerge to strike their chimes.

All of that apparatus is housed in a structure of unembarrassed eclecticism, both stylistic and intellectual. The central tower of the clock is topped with a froth of German baroque frosting, whereas the smaller turret on the left (which houses the weights that drive the clockwork) has been given a more Frenchified treatment. The third tower, on the right, is a stone spiral staircase that might have been salvaged from an Italian Renaissance belvedere. In the base of the cabinet, two glass panels allowing a view of brass gear trains are a distinctively nineteenth-century element; they look like the store windows of an apothecary's shop. The paintings and statues are mainly on religious themes — death and resurrection, fall and salvation — but they also include portraits of Urania (the muse of astronomy) and Copernicus. Another painting portrays Jean-Baptiste Schwilgué, whose part in this story I shall return to presently.

It's all done with gears. Also pinions, worms, snails, arbors; pawls and ratchets; cams and cam followers; cables, levers, bell cranks, and pivots.

The actual time-keeping mechanism — a pendulum and escapement much like the ones present in other clocks — drives the gear train for mean solar time. All the other astronomical and calendrical functions are derived from that basic, steady motion. For example, local solar time is calculated by applying two corrections to mean solar time. The first correction compensates for seasonal changes in the length of the day, the second for variations in the earth's orbital velocity as it follows its slightly elliptical path around the sun. The corrections are computed by a pair of "profile wheels," whose rims are machined to trace out a graph of the appropriate mathematical function. A roller, following the profile as the wheel turns, adjusts the speed of the local-solar-time pointer accordingly. The computation of lunar motion requires five correction terms and five profile wheels. They all have names: anomaly, evection, variation, annual equation, reduction.

The overall accuracy of the clock can be no better than the adjustment of the pendulum, which requires continual intervention, but for the subsidiary time-keeping functions there is another kind of error to be considered as well. Even if the mean solar time is exact, will all the solar, lunar, and planetary indicators keep pace correctly? The answer depends on how well celestial motions can be approximated by rational arithmetic — specifically, by gear ratios. The Strasbourg clock comes impressively close. For example, the true sidereal day is 23 hours, 56 minutes, 4.0905324 seconds, whereas the mean solar day is, by definition, exactly 24 hours. The ratio of the two intervals is 78,892,213 to 79,108,313, but grinding gears with nearly 80 million teeth is out of the question. The clock approximates the ratio as the reciprocal of $1 + (450/611 \times 1/269)$, which works out to a sidereal day of 23 hours, 56 minutes, 4.0905533 seconds. The error is less than a second a century.

The most intricate calculations are the ones for leap years and the movable feasts of the church. The rule for leap years states that a year N has an extra day if N is divisible by 4, unless N is also divisible by 100, in which case the year is a common year, with only the usual 365 days — but if N happens also to be divisible by 400, the year becomes a leap year again. Thus 1700, 1800, and 1900 were all common years, but 2000 will have a February 29. How can you encode such a nest of if-then-else rules in a gear train?

The clock has a wheel with twenty-four teeth and space for an omitted twenty-fifth. That wheel is driven at a rate of one turn per century, and so every four years a tooth comes into position to actuate the leap-year mechanism. The gap where the twenty-fifth tooth would be takes care of the divisible-by-100 exception. For the divisible-by-400 exception, a second wheel turns once every 400 years. It carries the missing twenty-fifth tooth and slides it into place on every fourth revolution of the century wheel, just in time to trigger the quadricentennial leap year.

The display of leap years calls for as much ingenuity as their calculation. On the large calendar ring, an open space between December 31 and January 1 bears the legend *Commencement de l'année commune* ("start of common year"). Shortly before midnight on the December 31 before a leap year, a sliding flange that carries the first sixty days of the year ratchets backward by the space of one day, covering up the word *commune* at one end of the flange and at the same time exposing February 29 at the other end. The flange remains in that position throughout the year, then shifts forward again to cover up the twenty-ninth and reveal *commune* just as the following year begins.

The rules for finding the date of Easter are even more intricate than the leap-year rule. Donald E. Knuth, in his *Art of Computer Programming*, remarks, "There are many indications that the sole important application of arithmetic in Europe during the Middle Ages was the calculation of [the] Easter date." Knuth's version of a sixteenth-century algorithm for the calculation has eight major steps, and some of the steps are fairly complicated. Here (to paraphrase the mathematics slightly) is step five:

> Calculate the sum $11G + 20 + Z - X$, where the numbers G, Z and X come from earlier steps in the algorithm. Now reduce that sum modulo 30 — that is, divide by 30 and keep only the remainder. Label the result E, for the so-called epact, the "age" of the moon at the start of the year. Finally, if E is equal to 25 and G is greater than 11, or if E is 24, then increase E by 1.

Programming a modern computer to perform the Easter calculation requires some care; programming a box of brass gears to do the arithmetic is truly a tour de force. I have stared at diagrams of the gears and linkages and tried to trace out their action, but I still don't fully understand how it all fits together.

In the abstract, it's not hard to see how a mechanical linkage could carry out the basic steps of the epact calculation. A wheel with thirty teeth or cogs would ratchet $11G$ notches clockwise, then add twenty steps more in the same direction, then another Z steps; finally, it would turn X steps counterclockwise. The "modulo 30" part of the program would be taken care of automatically if the arithmetic were done on a circle with thirty divisions. So far, so good. The thirty-tooth wheel does exist in the Strasbourg clock, and it is even helpfully labeled *Epacte*. Where I get lost is in trying to understand the various lever arms and rack-and-pinion assemblies that drive the epact wheel, and the cam followers that communicate its state to the rest of the system. There appear to be a number of optimizations in the works, which doubtless save a little brass but make the operation more obscure. Perhaps if I had a model of the clock that I could take apart and put back together . . .

But never mind my failures of spatiotemporal reasoning. The mechanism does work. Each New Year's Eve a metal tag that marks the date of Easter slides along the circumference of the calendar ring and takes up a position over the correct Sunday for the coming year. All the other movable feasts of the church are determined by the date of Easter, so the indicators of their dates are linked to the Easter tag and move along with it.

The present Strasbourg clock is the third in a series. The first was built in the middle of the fourteenth century, just as the cathedral itself was being completed. The original clock had three mechanical Magi that bowed down before the Virgin and child every hour on the hour.

By the middle of the sixteenth century, the Clock of the Three Kings was no longer running and no longer at the leading edge of horological technology. To supervise an upgrade, the Strasbourgeois hired Conrad Dasypodius, the professor of mathematics at Strasbourg, as well as the clock maker Isaac Habrecht and the artist Tobias Stimmer. Those three laid out the basic plan of the instrument still seen today, including the three-turreted case and most of the paintings and sculptures. A curiosity surviving from that era is the portrait of Copernicus — a curiosity because the planetary display on the Dasypodius clock was Ptolemaic. The second clock lasted another 200 years or so.

The story of the third clock starts with an anecdote so charming that I can't bear to look too closely into its authenticity. Early in the 1800s, the story goes, a beadle was giving a tour of the cathedral and mentioned that the clock had been stopped for twenty years. No one knew how to fix it. A small voice piped up: "*I will make it go!*" The boy who made the declaration was Jean-Baptiste Schwilgué, and forty years later he made good on his promise.

There was mild conflict over the terms of Schwilgué's commission. He wanted to build an entirely new clock; the cathedral administration wanted to repair the old one. They compromised: he gutted the works but kept the case, and built his new indicators and automata to fit the old design. The new mechanism began ticking on October 2, 1842.

Schwilgué was clearly thinking long-term when he undertook the project. As I noted earlier, the leap-year mechanism includes parts that engage only once every 400 years — parts that will soon be tested for the first time and then lie dormant again until 2400. Such very rare events might have been left for manual correction: it would have been only a small imposition on the clock's maintainers to ask that the hands be reset every four centuries. But Schwilgué evidently took pride and pleasure in getting the details right. He couldn't know whether the clock would still be running in 2000 or 2400, but he could build it in such a way that if it *did* survive, it would not perpetrate error.

The contrast with recent practice in computer hardware and software could hardly be more stark. Some computer programs, even if they survive the Y2K scare, are explicitly limited to dates between 1901 and 2099. The reason for choosing that particular span is that it makes the leap-year rule so simple: it's just a test of divisibility by 4. Under the circumstances, that design choice seems pretty wimpy. If Schwilgué could take the trouble to fabricate wheels that make one revolution every 100 years and every 400 years, surely a programmer could write the extra line of code needed to check for the century exceptions. The line might never be needed, but there's the satisfaction of knowing it's there.

Other parts of Schwilgué's clock look even further into the future. There is a gear deep in the works of the ecclesiastical computer that turns once every 2,500 years. And the celestial sphere out in front of the clock has a still slower motion. In addition to the sphere's daily rotation, it pirouettes slowly on another axis to reflect the precession of the equinoxes of the earth's orbit through the constellations of the zodiac. In the real solar system, that stately motion is what has lately brought us to the dawning of the Age of Aquarius. In the clock, the once-per-sidereal-day spinning of the globe is geared down at a ratio of 9,451,512 to 1, so that the equinoxes will complete one full precessional cycle after the passage of a bit more than 25,806 years. (The actual period is now thought to be 25,784 years.) At that point we'll be back to the cusp of Aquarius again, and no doubt paisley bell-bottoms will be back in fashion.

The odometer of years, as I mentioned earlier, runs to 9999. According to some accounts, Schwilgué suggested that if the clock is still going in 10000, a numeral *1* could be painted to the left of the thousands digit. The simplicity of that proposed solution suggests that the Y10K crisis may turn out to be less severe than the Y2K one.

Is there any chance the Strasbourg clock will actually run for 10,000 years? No products of human artifice have yet lasted so long, with the exception of cave paintings and some sharpened flints. Stonehenge and the pyramids of Egypt are half that age. The two earlier Strasbourg clocks, built with similar technology, both failed after roughly two centuries. Very few complex machines with moving parts have lasted more than a few hundred years.

Even if the clock keeps ticking, will anyone in 11999 want to know the date of Easter? Will people still be counting the years of the Common Era? No system of time-keeping has lasted anywhere near 10,000 years. The Roman calendar was abandoned after 1,500 years; the Mayan one may have lasted as long as 2,000 years; the Egyptian, possibly 3,000. The Chinese have been recording dates by cycles and reigns for something like 3,500 years. The Hebrew calendar is at the year 5760 — but that's not to say the scheme has been in use that long (there was no one around, after all, to turn the page on 1 Tishri 1). The Julian day system was invented only 400 years ago. Meanwhile, other calendars have come and gone. If Schwilgué had rebuilt the Strasbourg clock just a few decades earlier, it would have listed dates in Brumaire, Thermidor, Fructidor, and the other months decreed by the French Revolution in September 1792, and the clock's register of years would now be reading just 208.

I want to address another question. Even if a clock can be kept ticking, and even if the calendar it keeps retains some meaning, is the building of such multimillennial machines a good idea? I have my doubts, and they have been redoubled by a recent proposal to build another 10,000-year clock.

The new plan comes from Danny Hillis, the architect of the Connection Machine, an innovative and widely admired supercomputer of the 1980s (Hillis is now at the Walt Disney Corporation in Glendale, California). Together with several friends and colleagues, he has proposed building a clock described as "the world's slowest computer." The project is outlined in *The Clock of the Long Now*, a book by Stewart Brand, the instigator of the *Whole Earth Catalog*.

Technical details of the Long Now clock remain to be worked out, but the provisional design that Brand describes has a torsion pendulum (one that twists rather than swings) and a digital counter of pendulum oscillations instead of an analog gear train. Although the counter is digital, it is emphatically not electronic; Hillis's design uses mechanical wheels, pegs, and levers to count in binary notation.

The plan is to build several clocks of increasing grandeur. A prototype now under construction will be eight feet high. A twenty-foot model will be placed in a large city for ease of access, then a sixty-footer will be installed somewhere out in the desert for safekeeping. Here is one of Hillis's visions of how the big clock might be experienced:

> Imagine the clock is a series of rooms. In the first chamber is a large, slow pendulum. This is your heart beating, but slower. In the next chamber is a simple twenty-four-hour clock that goes around once a day. In the next chamber, just a Moon globe, showing the phase of the lunar month. In the next chamber is an armillary sphere tracking the equinoxes, the solstices, and the inclination of the Sun. . . .
> The final chamber is much larger than the rest. This is the calendar room. It contains a ring that rotates once a century and the 10,000-year segment of a much larger ring that rotates once every precession of the equinoxes. These two rings intersect to show the current calendrical date.

The motive for building such a monument to slow motion is not time-keeping per se; Hillis is not worried about losing count of the centuries. The aim is psychological. The clock is meant to encourage long-term thinking, to remind people of the needs and claims of future generations. The preamble to the project summary begins, "Civilization is revving itself into a pathologically short attention span. The trend might be coming from the acceleration of technology, the short-horizon perspective of market-driven economics, the next-election perspective of democracies, or the distractions of personal multitasking." The big, slow clock would offer a counterpoise to those frenetic tendencies; it would "embody deep time."

The wisdom of planning ahead, husbanding resources, saving something for those who will come after, leaving the world a better place — it's hard to quibble with all that. Concern for the welfare of one's children and grandchildren is surely a virtue — or at least a Darwinian imperative — and more general benevolence toward future inhabitants of the planet is also widely esteemed. But if looking ahead two or three generations is good, does that mean looking ahead twenty or thirty generations is better? What about 200 or 300 generations? Perhaps the answer depends on how far ahead you can actually see.

The Long Now group urges us to act in the best interests of posterity, but beyond a century or two I have no idea what those interests might be. To assume that the values of our own age embody eternal verities and virtues is foolish and arrogant. For all I know, some future generation will thank us for burning up all that noxious petroleum and curse us for exterminating the smallpox virus.

From a reading of Brand's book, I don't sense that the Long Now organizers can see any further ahead than the rest of us; as a matter of fact, they seem to be living in quite a short Now. All those afflictions listed in their preamble — the focus on quarterly earnings, quadrennial elections, and so forth — are bugaboos of recent years and decades. They would have been incomprehensible a few centuries ago, and there's not much reason to suppose they will make anybody's list of pressing concerns a few centuries hence.

The emphasis on the superiority of binary digital computing is something else that puts a late-twentieth-century date stamp on the project. A time may come when Hillis's binary counters will look just as quaint as Schwilgué's brass gears.

Long-term thinking is really hard. Of course, that's the point of the Long Now project, but it's also a point of weakness. It's hard to keep in mind that what seems most steadfast over the human life span may be evanescent on a geological or astronomical timescale. Consider the plan to put one clock in a city (New York, say) and another in a desert (Nevada). This makes sense now, but will New York remain urban and Nevada sparsely populated for the next 10,000 years? Many a desolate spot in the desert today was once a city, and vice versa.

Needless to say, the difficulty of predicting the future is no warrant to ignore it. The current Y2K predicament is clear evidence that a time horizon of two digits is too short. But four digits is plenty. If we take up the habit of building machines meant to last past the year 10000, or if we write our computer programs with room for five-digit years, we are not doing the future a favor. We're merely nourishing our own delusions. In the 1500s, Dasypodius and his colleagues could have chosen to restore the 200-year-old Clock of the Three Kings in Strasbourg Cathedral, but instead they ripped out all traces of it and built a new and better clock. Two hundred years later, Schwilgué was asked to repair the Dasypodius clock, but instead he eviscerated it and installed his own mechanism in the hollowed-out carcass. Today, after another two centuries, the Long Now group is not threatening to destroy the Schwilgué clock, but neither are they working to ensure its longevity. They ignore it. They want to build a newer, better, different clock, good for 10,000 years.

I begin to detect a pattern. The fact is, winding and dusting and fixing somebody else's old clock is boring. Building a brand-new clock of your own is much more fun, particularly if you can pretend that it's going to inspire awe and wonder for ages to come. So why not have the fun now and let the next 300 generations do the boring parts?

If I thought that Hillis and his associates might possibly succeed in this act of chronocolonialism, enslaving future generations to maintain our legacy systems, I would consider it my duty to posterity to oppose the project, even to sabotage it. But in fact I don't worry. I have faith in the future. Sometime in the 2100s a small child touring the ruins of the Clock of the Long Now will proclaim: "*I* will make it go!" And that child will surely scrap the whole mess and build a new and better clock, good for 10,000 years.

EDWARD HOAGLAND

That Sense of Falling

FROM *Preservation*

FOR ME, there is some vertigo in peering over the precipice of the new millennium. Although no expert, I see a kind of centrifugal seething if I venture to look a little bit over the edge (which most of us don't seem to enjoy doing). Density and yet attenuation. Medical science persists in elongating our life span and sharpening fertility techniques: Yet how will we crowd so many more people in? It's not just a matter of growing enough protein, but a spiritual question. If you multiply them endlessly, do human beings retain the sacred spark we have tried to ascribe to each of them? Is it likely, in other words, that we will continue to regard human life as special; that the human form won't come to look merely spidery or crabby and no longer a visual magnet? Will donor fatigue, so familiar from watching TV, expand exponentially to an empathy drain, and what we call road rage to a spatial exasperation that is far more pervasive? And if we don't feel special (I've never thought we were, except as part of a sacred whole), but only claustrophobic, what will be left? With a thinned-out populace of birds, few fish in the ocean, and a bare-bones terrestrial ecology, the honeybees, turtles, chimpanzees, butterflies, and what-have-you gone, will we feel skeletal ourselves? I think so.

But faith adjusts flexibly to avoid bafflement. God forbid we should ever feel surprised. And travel speeds up, so that as more and more appears banal close to home, we can touch down on other continents. With money, we can leave the world having visited a good deal of it. Generosity, however, may be more imperiled, now that people move a lot and depersonalize or monetarize so much of life. It's not even greed as much as digitry — paper debts and paper profits, and the facile ease with which new friendships can be formed or discarded in cyberspace. Mutuality is a mutual fund. Faith and sightseeing can be selfish, to suit an era when people live so often in communities segregated by income but not by loyalties of any sort. Will life be weightless like a Web site?

A loss of privacy won't be the least of it, or the worst, either. To mince up all the space we've had, dice, bleach, and colorize it, virtualize it into a simulacrum or simulation, severing ourselves from the tactile world, is an extraordinary, almost Martian experiment — giddy as Plato's shadowy cave, where only figments of the outside world were flashed upon the wall. How sudden a divorce from the acculturation of the past can we absorb? And how abrupt an amputation from the rest of nature can we stand before our nerves fray past repairing? How will the capacity for tact, a feel for body language, facial expression, and the value of compassion, be conveyed between the generations if education is computerized? Or a sense of the four seasons, the moon's phases, wind direction, meteorological rhythms, and being diurnal or nocturnal — of the sky as the heavens, not a periodic inconvenience — if we seldom leave our screens? What remains to be played out is whether the Internet and World Wide Web can be netlike, a true web, and catch us, or just a device for ventriloquism, spinning us outward, formless and flailing, juggling acquaintances, jamming phenomenal numbers of communications into a handful of hours while plucking news briefs, stock quotes, "spam" scam, business e-mail, family updates, and other automated miscellany from the ether or a wire.

Science is not sluggardly yet seems devoid of grief, because this would be a life with Mozart or other succulent choices at our fingertips, but oddly truncated, with so little sky and green and random sound or scent blowing in. We may need to grow not only hydroponic vitamins, but also oxygen, if the forests and oceanic vegetation are mauled beyond resuscitation: breathing units, to complement what may be denoted as affection units once the components of a child's emotional needs have been mapped precisely. I believe we will miss more than we imagine now the two-thirds of all species that are being vacuumed up in front of our juggernaut. Throbbing to some of the same tempos that we use, they lived in concert with us for so long, it would be illogical to suppose that we aren't going to feel lonely after they're gone. Having presumed that God gave us the earth to spend, we've positively splurged and, running through our inheritance, are already groping in some emptied pockets. In an amazing miscalculation, we seem to have thought that God was a sort of Aladdin's genie: If you rub him the right way, he will make more fish after we've sieved out the ocean; more topsoil if we've stripped it off and paved it over; more drinking water as we siphon dry the aquifers. Manlike, he will respond to pious stroking, and if the poles melt and the oceans brim, he'll throw us a rope, build us an ark.

Of course this is nonsense — but has been our assumption. The famines, water shortages, and weather eruptions will be episodic and not unique to history except in their ubiquity. But what is changing in the meantime is our relationship to one another. The links that enable people to ride out disasters, hand in hand or eye to eye, knowing the commonalities that they hold dear, and quite specifically what they fear, are being snipped off. What is precious now? Having children is again in, yet to nurture them toward what? Ethnic hatred is out, but a negative is not enough. People have become so solipsistic as well as secular that they aren't allied, affiliated, or even rooted as to place. They want money, status, but not the allegiances that used to accompany these. Diffusion is the rule, with serial marriages, steppingstone suburbs, segmented careers. And if there is a precipice, it will be a bumping tumble that we are in for, where the rolling body hits assorted ledges and momentary sanctuaries, scrabbling sporadically to hold on all the way down.

Personally, I live for love — love of the world — indeed, immersion in it. But what is confusing is that the world I love is being destroyed, and I can't simply play the codger and say that everything is going to hell since I've turned sixty, because I notice as a college teacher, for instance, that students are as idealistic and intelligent as ever. What's dissolving is the framework they will operate in. And we can't wire it back together: not the cultural constellations and not nature's infinite anatomy. Wires are only wires. In simulating, they sometimes blow a fuse or can be flicked off with a switch. Both culture and nature have counted because they don't and can't. Rather, at our instigation they can get delirious and sick and do us damage with guns gone haywire or tornadoes. We have been laying

waste to so much so fast, perhaps figuring that we could store its im-
agery on disks, the effects already are incalculable. I have that
sense of falling.

A farmer living near me once got so fed up that he turned his
tractor off the road and drove it straight through the side of his
own barn. You can still see the hole. We're not helpless. We can live
privately; draw in our wagons or Web sites; remodel rambling old
houses instead of jimmying new ones into gridded spaces; read
good books and be ironic about progress, expressing an interest in
space travel for warehousing excess human beings. I put out bird-
baths and bird feeders and am moving a local colony of garter
snakes from in front of a road crew's bulldozers before the asphalt
seals them underground. Yet the solutions and solace are temporary, like wiring out uncomfortable information. Events will take
their course and end the same.

JUDITH HOOPER

A New Germ Theory

FROM *The Atlantic Monthly*

A LATE-SEPTEMBER heat wave enveloped Amherst College, and young people milled about in shorts or sleeveless summer frocks, or read books on the grass. Inside the red-brick buildings framing the leafy quadrangle, students listened to lectures on Ellison and Emerson, on Paul Verlaine and the Holy Roman Empire. Few suspected that strains of the organism that causes cholera were growing nearby, in the Life Sciences Building. If they had known, they would probably not have grasped the implications. But these particular strains of cholera make Paul Ewald smile; they are strong evidence that he is on the right track. Knowing the rules of evolutionary biology, he believes, can change the course of infectious disease.

In a hallway of the Life Sciences Building, an anonymous student has scrawled above a display of glossy photographs and vitae of the faculty, "We are the water; you are but the sponge." This is the home of Amherst's biology department, where Paul Ewald is a professor. He is also the author of the seminal book *Evolution of Infectious Disease* and of a long list of influential papers. Sandy-haired, trim, and handsome in an all-American way, he looks considerably younger than his forty-five years. Conspicuously outdoorsy for an academic, he would not seem out of place in an L. L. Bean catalogue, with a golden retriever by his side. Ewald rides his bike to the campus every day in decent weather — and in weather one might not consider decent — from the nearby hill village of Shutesbury, where he lives with his wife, Chris, and two teenage children in a restored eighteenth-century house.

As far as Ewald is concerned, Darwin's legacy is the most interesting thing on the planet. The appeal of evolutionary theory is that it is a grand unifying principle, linking all organisms, from protozoa to presidents, and yet its essence is simple and transparent. "Darwin only had a couple of basic tenets," Ewald observed recently in his office. "You have heritable variation, and you've got differences in survival and reproduction among the variants. That's the beauty of it. It has to be true — it's like arithmetic. And if there is life on other planets, natural selection has to be the fundamental organizing principle there, too."

These Darwinian laws have led Ewald to a new theory: that diseases we have long ascribed to genetic or environmental factors — including some forms of heart disease, cancer, and mental illness — are in many cases actually caused by infections. Before we take up this theory, we need to spend a moment with Ewald's earlier work.

Ewald began in typical evolutionary terrain, studying hummingbirds and other creatures visible to the naked eye. It was on a 1977 field trip to study a species called Harris's sparrow in Kansas that a bad case of diarrhea laid him up for a few days and changed the course of his career. The more he meditated on how Darwinian principles might apply to the organisms responsible for his distress — asking himself, for instance, what impact treating the diarrhea would have on the vast populations of bacteria evolving within his intestine — the more obsessed he became. Was his diarrhea a strategy used by the pathogen to spread itself, he wondered, or was it a defense employed by the host — his body — to flush out the invader? If he curbed the diarrhea with medication, would he be benefiting the invader or the host? Ewald's paper outlining his speculations about diarrhea was published in 1980, in the *Journal of Theoretical Biology*. By then Ewald was on his way to becoming the Darwin of the microworld.

"Ironically," he says, "natural selection was first recognized as operating in large organisms, and ignored in the very organisms in which it is especially powerful — the microorganisms that cause disease. The timescale is so much shorter and the selective pressures so much more intense. You can get evolutionary change in disease organisms in months or weeks. In something like zebras you'd have to wait many centuries to see it."

For decades, medical science was dominated by the doctrine of "commensalism" — the notion that the pathogen-host relationship inevitably evolves toward peaceful coexistence, and the pathogen itself toward mildness, because it is in the germ's interest to keep its host alive. This sounds plausible, but it happens to be wrong. The Darwinian struggle of people and germs is not necessarily so benign. Evolutionary change in germs can go either way, as parasitologists and population geneticists have realized — toward mildness or toward virulence. It was Ewald's insight to realize what we might do about it.

Manipulating the Enemy

Say you're a disease organism — a rhinovirus, perhaps, the cause of one of the many varieties of the common cold; or the mycobacterium that causes tuberculosis; or perhaps the pathogen that immobilized Ewald with diarrhea. Your best bet is to multiply inside your host as fast as you can. However, if you produce too many copies of yourself, you'll risk killing or immobilizing your host before you can spread. If you're the average airborne respiratory virus, it's best if your host is well enough to go to work and sneeze on people in the subway.

Now imagine that host mobility is unnecessary for transmission. If you're a germ that can travel from person to person by way of a "vector," or carrier, such as a mosquito or a tsetse fly, you can afford to become very harmful. This is why, Ewald argues, insect-borne diseases such as yellow fever, malaria, and sleeping sickness get so ugly. Cholera uses another kind of vector for transmission: it is generally waterborne, traveling easily by way of fecal matter shed into the water supply. And it, too, is very ugly.

"Here's the [safety] hood where we handle the cholera," Jill Saunders explained as we toured the basement lab in Amherst's Life Sciences Building where cholera strains are stored in industrial refrigerators after their arrival from hospitals in Peru, Chile, and Guatemala. "We always wear gloves." A medical-school-bound senior from the Boston suburbs, Saunders is one of Ewald's honor students. As she guided me around, pointing out centrifuges, −80 degree freezers, and doors with BIOHAZARD warnings, we passed a closet-sized room as hot and steamy as the tropical zones where hemorrhagic fevers thrive. She said, "This is the incubation room, where we grow the cholera."

Cholera invaded Peru in 1991 and quickly spread throughout South and Central America, in the process providing a ready-made experiment for Ewald. On the day of my tour, Saunders had presented to the assembled biology department her honors project, "Geographical Variations in the Virulence of *Vibrio cholerae* in Latin America." The data compressed in her tables and bar graphs were evidence for Ewald's central thesis: it is possible to influence a disease organism's evolution to your advantage. Saunders used a standard assay, called ELISA, to measure the amount of toxin produced by different strains of cholera, thus inferring the virulence of *V. cholerae* variants from several Latin American regions. Then she and Ewald looked at figures for water quality — what percentage of the population had potable water, for example — and looked for correlations. If virulent strains correlated with a contaminated water supply, and if, conversely, mild strains took over where the water was clean, the implication would be that *V. cholerae* becomes increasingly mild when it cannot use water as a vector. When the pathogen is denied easy access to new hosts through fecal matter in the water system, its transmission depends on infected people moving into contact with healthy ones. In this scenario the less-toxic variants would prevail, because these strains do not incapacitate or kill the host before they can be spread to others. If this turned out to be true, it would constitute the kind of evidence that Ewald expected to find.

The dots on Saunders's graphs made it plain that cholera strains are virulent in Guatemala, where the water is bad, and mild in Chile, where water quality is good. "The Chilean data show how quickly it can become mild in response to different selective pressures," Ewald explained. "Public-health people try to keep a disease from spreading in a population, and they don't realize that we can also change the organism itself. If you can make an organism very mild, it works like a natural vaccine against the virulent strains. That's the most preventive of preventive medicine: when you can change the organism so it doesn't make you sick." Strains of the cholera agent isolated from Texas and Louisiana produce such small amounts of toxin that almost no one who is infected with them will come down with cholera.

Joseph Schall, a professor of biology at the University of Vermont, offers a comment on Ewald's work: "If Paul is right, it may be that the application of an evolutionary theory to public health could save millions of lives. It's a stunning idea. If we're able to manipulate the evolutionary trajectory of our friends — domestic animals and crops — why not do the same with our enemies, with cholera, malaria, and HIV? As Thomas Huxley said when he read Darwin, 'How stupid of me not to have thought of that before.' I thought when I heard Paul's idea, 'Gee, why didn't *I* think of that?'"

Ewald put forward his virulence theories in *Evolution of Infectious Disease*. Today his book is on the syllabus for just about every college course in Darwinian medicine or its equivalent. "I regard him as a major figure in the field," says Robert Trivers, a prominent evolutionary biologist who holds professorships in anthropology and biology at Rutgers University. "It is a shame his work isn't better known to the public-health and medical establishments, who are willfully ignorant of evolutionary logic throughout their training." While praising Ewald's boldness and originality, some of his peers caution that his data need to be independently corroborated, and others object that his hypotheses are too crude to capture the teeming complexity of microbial evolution. "Evolutionary biologists have had very poor success in explaining how an organism evolves in response to its environment," says James Bull, an evolutionary geneticist at the University of Texas. "Trying to understand a two-species interaction should be even more complicated."

Recently, in any case, Ewald has adopted a new cause, far more radical but equally rooted in evolution. Let's call it Germ Theory, Part II. It offers a new way to think about the causes of some of humanity's chronic and most baffling illnesses. Ewald's view, to put it simply, is that the culprits will often turn out to be pathogens — that the dictates of evolution virtually demand that this be so.

The Case for Infection

Germ Theory, Part I, the edifice built by men like Louis Pasteur, Edward Jenner, and Robert Koch, took medicine out of the Dark Ages. It wasn't "bad air" or "bad blood" that caused diseases like malaria and yellow fever but pathogens transmitted by mosquitoes.

Tuberculosis was famously tracked to an airborne pathogen, *Myco-bacterium tuberculosis,* by Robert Koch, the great German scientist who in 1905 won a Nobel Prize for his work. Koch also revolution-ized medical epidemiology by laying out his famous four postu-lates, which have set the standard for proof of infectivity up to the present day. The postulates dictate that a microbe must be (a) found in an animal (or person) with the disease, (b) isolated and grown in culture, (c) injected into a healthy experimental animal, producing the disease in question, and then (d) recovered from the experimentally diseased animal and shown to be the same pathogen as the original.

By the early twentieth century the whole landscape had changed. Most of the common killer diseases, including smallpox, diphtheria, bubonic plague, flu, whooping cough, yellow fever, and TB, were understood to be caused by pathogens. Vaccines were de-vised against some, and by the 1950s antibiotics could easily cure many others. Smallpox was actually wiped off the face of the earth (if you don't count a few strains preserved in laboratories in the United States and Russia).

By the 1960s and 1970s the prevailing mood was one of opti-mism. Ewald is fond of quoting from a 1972 edition of a classic medical textbook: "The most likely forecast about the future of in-fectious disease is that it will be very dull." At least in the developed world, infectious diseases no longer seemed very threatening. Far scarier were the diseases that the medical world said were not infec-tious: heart disease, cancer, diabetes, and so on. No one foresaw the devastation of AIDS, or the serial outbreaks of deadly new in-fections such as Legionnaires' disease, Ebola and Marburg hemor-rhagic fevers, antibiotic-resistant tuberculosis, "flesh-eating" staph infections, hepatitis C, and Rift Valley fever.

The infectious age is, we now know, far from over. Furthermore, it appears that many diseases we didn't think were infectious may be caused by infectious agents after all. Ewald observes, "By guid-ing researchers down one path, Koch's postulates directed them away from alternate ones. Researchers were guided away from dis-eases that might have been infectious but had little chance of ful-filling the postulates." That is, just because we couldn't readily dis-cover their cause, we rather arbitrarily decided that the so-called chronic diseases of the late twentieth century must be hereditary or environmental or "multifactorial." And, Ewald contends, we have frequently been wrong.

Germ Theory, Part II, as conceived by Ewald and his collaborator, Gregory M. Cochran, flows from the timeless logic of evolutionary fitness. Coined by Darwin to refer to the fit between an organism and its environment, the term has come to mean the evolutionary success of an organism relative to competing organisms. Genetic traits that may be unfavorable to an organism's survival or reproduction do not persist in the gene pool for very long. Natural selection, by its very definition, weeds them out in short order. By this logic, any inherited disease or trait that has a serious impact on fitness must fade over time, because the genes that spell out that disease or trait will be passed on to fewer and fewer individuals in future generations. Therefore, in considering common illnesses with severe fitness costs, we may presume that they are unlikely to have a genetic cause. If we cannot track them to some hostile environmental element (including lifestyle), Ewald argues, then we must look elsewhere for the explanation. "When diseases have been present in human populations for many generations and still have a substantial negative impact on people's fitness," he says, "they are likely to have infectious causes."

Although its fitness-reducing dimensions are difficult to calculate, the ordinary stomach ulcer is the best recent example of a common ailment for which an infectious agent — to the surprise of almost everyone — turns out to be responsible.

When I visited him one afternoon, Ewald pulled off his shelves a standard medical textbook from the 1970s and opened the heavy volume to the entry on peptic ulcers. We squinted together at a gray field of small print punctuated by subheads in boldface. Under "Etiology" we scanned several pages: *environmental factors . . . smoking . . . diet . . . ulcers caused by drugs . . . aspirin . . . psychonomic factors . . . lesions caused by stress*. In the omniscient tone of medical texts, the authors concluded, "It is plausible to hypothesize a wealth of these factors. . . ." There was no mention of infection at all.

In 1981 Barry J. Marshall was training in internal medicine at the Royal Perth Hospital, in Western Australia, when he became interested in incidences of spiral bacteria in the stomach lining. The bacteria were assumed to be irrelevant to ulcer pathology, but Marshall and J. R. Warren, a histopathologist who had previously observed the bacteria, reviewed the records of patients whose stomachs were infected with large numbers of these bacteria. They noticed that when one patient was treated with tetracycline for unrelated reasons, his pain vanished, and an endoscopy revealed that his ulcer was gone.

An article by Marshall and Warren on their culturing of "unidentified curved bacilli" appeared in the British medical journal *The Lancet* in 1984, and was followed by other suggestive studies. For years, however, the medical establishment remained deaf to their findings, and around the world ulcer patients continued to dine on bland food, swear off stress, and swill Pepto-Bismol. Finally, Marshall personally ingested a batch of the spiral bacteria and came down with painful gastritis, thereby fulfilling all of Koch's postulates.

There is now little doubt that *Helicobacter pylori*, found in the stomachs of a third of adults in the United States, causes inflammation of the stomach lining. In 20 percent of infected people it produces an ulcer. Nearly everyone with a duodenal ulcer is infected. *H. pylori* infections can be readily diagnosed with endoscopic biopsy tests, a blood test for antibodies, or a breath test. In 90 percent of cases the infections can be cured in less than a month with antibiotics. (Unfortunately, many doctors still haven't gotten the news. A Colorado survey found that 46 percent of patients seeking medical attention for ulcer symptoms are never tested for *H. pylori* by their physicians.)

Antibiotics Against Heart Disease?

Ewald closed the medical textbook on his knee. "This was published twenty years ago," he said. "If we looked up 'atherosclerosis' in a textbook from ten years ago, we'd find the same kind of things — stress, lifestyle, lots about diet, nothing about infection."

Heart disease is now being linked to *Chlamydia pneumoniae*, a newly discovered bacterium that causes pneumonia and bronchitis. The germ is a relative of *Chlamydia trachomatis*, which causes trachoma, a leading cause of blindness in parts of the Third World. *C. trachomatis* is perhaps more familiar to us as a sexually transmitted disease that, left untreated in women, can lead to scarring of the fallopian tubes, pelvic inflammatory disease, ectopic pregnancy, and tubal infertility.

Pekka Saikku and Maija Leinonen, a Finnish husband-and-wife team who have evoked comparisons to the Curies, discovered the new type of chlamydial infection in 1985, though its existence was not officially recognized until 1989. Saikku and Leinonen found that 68 percent of Finnish patients who had suffered heart attacks had high levels of antibodies to *C. pneumoniae*, as did 50 percent of patients with coronary artery disease, in contrast to 17 percent of the healthy controls. "We were mostly ignored or laughed at," Saikku recalls.

While examining coronary artery tissues at autopsy in 1991, Allan Shor, a pathologist in Johannesburg, saw "pear-shaped bodies" that looked like nothing he'd ever seen before. He mentioned his observations to a microbiologist colleague, who had read about a new species of chlamydia with a peculiar pear shape. The colleague referred Shor to an expert on the subject, Cho-Chou Kuo, of the University of Washington School of Public Health, in Seattle. After Shor shipped Kuo the curious coronary tissue, Kuo found that the clogged coronary arteries were full of *C. pneumoniae*. Before long, others were reporting the presence of live *C. pneumoniae* in arterial plaque fresh from operating tables. Everywhere the bacterium lodges, it appears to precipitate the same grim sequence of events: a chronic inflammation, followed by a buildup of plaque that occludes the opening of the artery (or, in the case of venereal chlamydia, a buildup of scar tissue in the fallopian tube). Recently a team of pathologists at MCP-Hahnemann School of Medicine, in Philadelphia, found the same bacterium in the diseased sections of the autopsied brains of patients who had had late-onset Alzheimer's disease: it was present in seventeen of nineteen Alzheimer's patients and in only one of nineteen controls.

By the mid-1990s a radical new view was emerging of atherosclerosis as a chronic, lifelong arterial infection. "I am confident that this will reach the level of certainty of ulcer and *H. pylori*," says Saikku, who estimates that at least 80 percent of all coronary artery disease is caused by the bacterium. Big questions remain, of course. Studies show that about 50 percent of U.S. adults carry antibodies to *C. pneumoniae* — but how many will develop heart disease? Even if heart patients can be shown to have antibodies to *C. pneumoniae*, and even if colonies of the bacteria are found living and breeding in diseased coronary arteries, is it certain that the germ *caused* the damage? Perhaps it's an innocent bystander, as some critics have proposed; or a secondary, opportunistic infection.

But suppose that a *Chlamydia pneumoniae* infection during childhood can initiate a silent, chronic infection of the coronary arteries, resulting in a "cardiovascular event" fifty years later. Could antibiotics help to address the problem?

A few early studies suggest they might. Researchers in Salt Lake City infected white rabbits with *C. pneumoniae,* fed them a modestly cholesterol-enhanced diet, killed them, and found thickening of the thoracic aortas, in contrast to the condition of uninfected controls fed the same diet. Additionally, treatment of infected rabbits with antibiotics in the weeks following infection prevented the thickening. Saikku and colleagues reported a similar finding, also in rabbits. Coronary patients in Europe who were treated with azithromycin not only showed a decline in antibodies and other markers of infection but in some studies had fewer subsequent cardiovascular events than patients who were given placebos. (These findings are preliminary; in a few years we may know more. The first major clinical trial is under way in the United States, sponsored by the National Institutes of Health and the Pfizer Corporation: 4,000 heart patients at twenty-seven clinical centers will be given either the antibiotic azithromycin or a placebo and followed for four years to gauge whether the antibiotic affects the incidence of further coronary events.)

Smoking, stress, cholesterol, and heredity all play a role in heart disease. But imagine if our No. 1 killer — with its vast culture of stress-reduction theories, low-fat diets, high-fiber cereals, cholesterol-lowering drugs, and high-tech bypass surgery — could in many instances be vanquished with an antibiotic. Numerous precedents exist for long-smoldering bacterial infections with consequences that appear months or years later. Lyme disease, leprosy, tuberculosis, and ulcers have a similar course. Ewald is confident that the association of *C. pneumoniae* and heart disease is real. He doesn't believe that the germ is an innocent bystander. "It reminds you a lot of gonorrhea in the 1890s," he says. "When they saw the organism there, people said, 'Well, we don't know if it's really causing the disease, or is just living there.' Every month the data are getting stronger. This is a smoking gun, just like *Helicobacter.*"

Evolutionary Byways

"I have a motto," Gregory Cochran told me recently. "'Big old diseases are infectious.' If it's common, higher than one in a thousand, I get suspicious. And if it's old, if it has been around for a while, I get suspicious."

The fact that Ewald has dared to conceive of a big theory for the medical sciences owes much to Cochran's contributions. A forty-five-year-old Ph.D. physicist who lives in Albuquerque with his wife and three small children, Cochran makes a living doing contract work on advanced optical systems for weaponry and other devices. Whereas Ewald is an academic insider, with department meetings to attend and honors theses to monitor, Cochran is a solo player, with an encyclopedic mind (he is a former College Bowl contestant) and a manner that verges on edginess. These days he spends a lot of time at his computer, as rapt as a conspiracy theorist, cruising Medline for new data on infectious diseases and, one imagines, almost cackling to himself when he finds something really good. Cochran's background in a field dominated by grand theories and universal laws may serve as a valuable counterpoint to the empirical and theory-hostile universe of the health sciences.

Ewald and Cochran encountered each other serendipitously, after Cochran decided to pursue a certain line of thinking about a very sensitive subject. "I was reading an article in *Scientific American* in 1992 about pathogens manipulating a host to get what they want," Cochran recalls. "It described a flowering plant infected by a fungus, and the fungus hijacks the plant's reproductive machinery so that instead of pollen it produces fungal spores. I thought, *Could it be?*" Cochran strayed from his field to try his hand at writing an article on biology — elaborating an audacious theory that human homosexuality might result from a "manipulation" of a host by a germ with its own agenda. He sent his draft to a prestigious biology journal, which sent it out to three scientists for peer review. Two were unconvinced, even appalled; the third was Paul Ewald, who thought the article was flawed but who was nonetheless impressed by the logic of the idea. The article was rejected, but Ewald and Cochran began their association.

To illustrate his thinking about infectiousness and disease, Cochran not long ago gave me a tour of his conceptual bins, into which he sorts afflictions according to their fitness impact. Remember that fitness can be defined as the evolutionary success of one organism relative to competing organisms. Only one thing counts: getting one's genes into the future. Any disease that kills host organisms before they can reproduce reduces fitness to zero. Obviously, fitness takes a major hit whenever the reproductive system itself is involved, as in the case of venereal chlamydia.

Consider a disease with a fitness cost of 1 percent — that is, a disease that takes a toll on survival or reproduction such that people who have it end up with 1 percent fewer offspring, on average, than the general population. That small amount adds up. If you have an inherited disease with a 1 percent fitness cost, in the next generation there will be 99 percent of the original number in the gene pool. Eventually the number of people with the disease will dwindle to close to zero — or, more precisely, to the rate produced by random genetic mutations: about 1 in 50,000 to 1 in 100,000.

We were considering the bin containing diseases that are profoundly antagonistic to fitness, with a fitness cost of somewhere between 1 and 10 percent by Cochran's calculations. My eye took in a catalogue of human ills — some familiar, some exotic, some historically fearsome but close to extinct, some lethal in the tropics but of little concern to inhabitants of the temperate zones. This list also showed prevailing medical opinion about cause. Each name of a disease was trailed by a lowercase letter: i (for infectious), g (genetic), $g+$ (genetic defense against an infectious disease), e (caused by an environmental agent), or u (unknown). I read, "Atherosclerosis (u), . . . chlamydia (venereal) (i), cholera (i), diphtheria (i), endometriosis (u), filariasis (i), G6PD deficiency $(g+)$, . . . hemoglobin E disease $(g+)$, hepatitis B (i), hepatitis C (i), hookworm disease (i), kuru (i), . . . malaria (vivax) (i), . . . pertussis (i), pneumococcal pneumonia (i), polycystic ovary disease (u), scarlet fever (i), . . . tuberculosis (i), typhoid (i), yellow fever (i)."

Of the top forty fitness-antagonistic diseases on the list, thirty-three are known to be directly infectious and three are indirectly caused by infection; Cochran believes that the others will turn out to be infectious too. The most fitness-antagonistic diseases must be infectious, not genetic, Ewald and Cochran reason, because otherwise their frequency would have sunk to the level of random mutations. The exceptions would be either diseases that could be the effect of some new environmental factor (radiation or smoking, for example), or genetic diseases that balance their fitness cost with a benefit. Sickle-cell anemia is one example of the latter.

Though sickle-cell anemia is strictly heritable according to Mendelian laws, it is widely believed to have persisted in the population in response to infectious selective pressures. It heads the list of genetic diseases that Ewald dubs "self-destructive defenses," in which a disease fatal in its homozygous form (two copies of the gene) carries an evolutionary advantage to heterozygous carriers (with one copy), protecting against a terrible infection: in this case falciparum malaria, common in Africa. Similarly, cystic fibrosis, some argue, evolved in northern Europe as a defense against *Salmonella typhi*, the cause of typhoid fever. Infection thus explains why these deadly genetic diseases have remained in the human gene pool when they should have died out.

But what about something like atherosclerosis? I asked. Leaving aside the evidence concerning *C. pneumoniae*, it is not apparent why a genetic cause for atherosclerosis should be dismissed out of hand on evolutionary grounds. If it hits people in midlife or later, after they have launched their genes, how could it possibly affect fitness?

Cochran's response illustrates some of the intricacies of evolutionary thinking. "Well, obviously, it's not as bad as a disease that kills you before puberty, but I think it does have a fitness cost. First of all, it's *really* common. Second, people think that all you have to do to pass your genes along is have children, but that's not true. You still need to raise the offspring to adulthood. In a hunter-gatherer or subsistence-farming culture, the fitness impact of dying in midlife might be considerable, especially during bad times, like famines. You've got to feed your family. Also, cardiovascular disease is a leading cause of impotence, and any disease that makes males impotent at age forty-five has got to affect reproduction somewhat."

But fifty-year-olds? Sixty-year-olds?

Grandmothers do a large proportion of the food gathering in some tribal cultures, according to recent anthropological reports. "They aren't hampered by babies anymore, and they don't have to go around chucking spears like the men," says George C. Williams, a professor emeritus of ecology and evolution at the State University of New York at Stony Brook, and one of the pillars of modern evolutionary biology. "They contribute substantially to the family diet." If long-lived elders historically have made a difference by fostering the survival of their descendants, and therefore their genes, Cochran figures, then a disease that kills sixty-year-olds could have a fitness impact of around 1 percent.

The Cause-and-Effect Conundrum

"Know what that is?" Ewald asked. We were standing in the main corridor of the Life Sciences Building, gazing up at a decorative metalwork frieze that runs along the walls just above door height. A pair of hummingbirds chase each other in a circle. A human eye and an octopus eye face off. A human hand is juxtaposed with a chimpanzee hand. Ewald pointed to something that looked like a daddy longlegs with a video camera for a head. "Some kind of insect?" I ventured. "It's a virus," he said. "See, it's like a spaceship. That" — he pointed at the head — "is its DNA. It injects it inside the cell."

There is something unsettling and fascinating about a virus, an organism that is neither strictly alive nor strictly inanimate, and that replicates by sneaking inside a host cell and commandeering its machinery. "Viruses are essentially bits of nucleic acid — either DNA or RNA — wrapped in a protein capsule," Ewald explained. "A retrovirus, like HIV, is an RNA virus with a protein called reverse transcriptase built into it, and once it gets into a cell, it uses the reverse transcriptase to make a DNA copy of its RNA. This viral DNA copy can insert itself into our DNA, where it can be read by our protein-making machinery the same way our own instructions are read."

The modus operandi of the world's most feared virus, HIV, is clever, killing its hosts very, very slowly. A sexually transmitted pathogen, without the luxury of being spread through sneezes or coughs, must await its few opportunities patiently; if those infected have no symptoms and don't know they are sick, so much the better. A mild, chronic form of AIDS had in all likelihood been around for centuries in Africa, according to Ewald. Suddenly in the 1970s — owing to changing patterns of sexual activity and to population movements — deadly strains spread in the population of central and East Africa.

HIV has an extremely high mutation rate, which means that it is continually evolving, even within a single patient, producing competing strains that fight for survival against the weapons produced by the immune system. If selective pressures — in this case a high potential sexual transmission — have forced the virus to evolve toward virulence, the opposite selective pressures could do the reverse. Conceivably, we could "tame" HIV, encouraging it to evolve toward comparative harmlessness. It was already known that preventive measures such as safe sex, fewer partners, clean needles, and so forth could curb the spread of the disease. But Ewald pointed out early on that social modification was a far more potent weapon than anyone realized. Once HIV was cut off from easy access to new hosts, milder strains would flourish — ones that the host could tolerate for longer and longer periods. Indeed, Ewald argues, given limited public-health budgets, it might make sense to put more money into transmission-prevention programs and less into the search for vaccines. (He also has strong opinions about how drugs should be used to treat AIDS. He asserts that every time we use an antiviral drug like AZT, we produce an array of AZT-resistant HIVs in the population; if viral evolution is taken into account, antiviral drugs can be used more judiciously.)

Ewald's theories tilt him decidedly toward the optimistic camp. Even in the absence of a vaccine, the AIDS epidemic will not inevitably worsen; it can be curbed without reducing transmission to zero. A natural experiment now occurring in Japan, he says, could be a test case for his theories. In the early 1990s highly virulent strains of HIV from Thailand took root in Japan, but Ewald predicts that low rates of sexual transmission in that country — due to widespread condom use and other factors — will act as a selective pressure on these strains, so that they evolve toward mildness. If this is true, the trend should become evident over the next ten years.

Like HIV, many other viruses have an indolent course, with a long latency between infection and the development of symptoms. Herpes zoster, the agent of chicken pox, lingers in the body forever, capable of erupting as painful shingles decades later. There are also so-called hit-and-run infections, in which a pathogen or its products disrupt the body's immunological surveillance system; once the microbes are gone (or when they are present in such low frequency as to be undetectable), the immune response stays stuck in the "on" mode, causing a lingering inflammation. By the time symptoms occur, the microorganism itself has disappeared, and its genome will not be detectable in any tissue.

"The health sciences are still grappling with the masking effects of long delays between the onset of infection and the onset of disease," Ewald says. "Any time you have hit-and-run infections, slow viruses, lingering or relapsing infections, or a time lag between infection and symptoms, the cause and effect is going to be very cryptic. You won't find these newly recognized infections by the methods we used to find old infectious diseases. We have to be ready to think of all sorts of new, clever ways to identify pathogens. We will have to abandon Koch's postulates in some cases."

The Great Synthesizer

As of this writing, the ideas at the core of Germ Theory, Part II, have been presented by Ewald mostly in the form of lectures, and in communications with colleagues. The papers in which the ideas will be formally articulated are in preparation. Given Ewald's prominence, the ideas are bound to cause a stir. They will also draw criticism. In the medical sciences, where "theory" is a bad word and "Stick to the data" is the reigning motto, Ewald will come under particular scrutiny because his hypothesis arrives detached from a vast corpus of laboratory data. It is helpful to think of Ewald as continuing the tradition of the great scientific synthesizers. Darwin himself was a synthesizer extraordinaire, who composed the thesis of *The Origin of Species* largely out of hundreds of odds and ends contributed by others, from pigeon breeders to naturalists. "Professor So-and-so has observed . . ." is a recurring motif in Darwin's book.

Ewald's theory about evolution and infectiousness provides a framework that potentially unites diverse research on the front lines of various afflictions. Ulcers and heart disease have already been mentioned. Here are two more: cancer and mental illness.

In 1910 a man named Peyton Rous discovered the eponymous Rous sarcoma virus, demonstrating that chickens infected with it developed cancer. Over the years many other cancer viruses have been discovered in animals. And yet until 1979, despite broad hints from the animal world, not a single human cancer was generally accepted as infectious. Rous had been lucky: his chickens became sick only two weeks after infection. Human cancers follow a more languorous course, which means that by the time symptoms show up, any infectious causation may well be buried under a lifetime of irrelevant risk factors.

In 1979, HTLV-1, a retrovirus endemic in parts of Asia, Africa, and the Caribbean, and transmitted either sexually or from mother to child, was linked to certain leukemias and lymphomas; the cancer appeared decades after infection. The Epstein-Barr virus (the agent that causes mononucleosis) has now been associated with some B-cell lymphomas, with a nasopharyngeal cancer common in south China, and with Burkitt's lymphoma, a deadly childhood cancer of Africa. Some 82 percent of all cases of cervical cancer have been associated with the sexually transmitted human papilloma virus, a once relatively innocent-seeming pathogen responsible for genital warts.

H. pylori, the ulcer pathogen, confers a sixfold greater risk of stomach cancer, and accounts for at least half of all stomach cancers. Also, the lymphoid tissue of the stomach can produce a low-grade gastric lymphoma under the influence of this bacterium. Early reports indicate that the lymphoma is cured in 50 percent of cases by resolving the *H. pylori* infection — which may mark the first time in medical history that cancer has been cured with an antibiotic.

Hepatitis B and C, two of the ever-growing alphabet soup of hepatic diseases, have been linked to liver cancer. Herpes virus 8 has recently been discovered to be the cause of Kaposi's sarcoma. "There is no reason to believe that this flurry of discovery has now completed the list of infectious agents of cancer," Ewald says.

Among the many known animal cancer viruses is a closely studied retrovirus known as mouse mammary tumor virus (MMTV), which causes mammary-gland cancer in mice. This virus is transmitted from mother to offspring through mother's milk, lying latent in the daughter's mammary tissue until activated by hormones during her own lactation. Could such a virus be a factor in human breast cancer? In the mid-1980s researchers announced that they had found in malignant human breast tumors a DNA sequence resembling MMTV, but the excitement waned when the same sequence was found in normal breast tissue as well. Interest has been revived by the research of Beatriz G.-T. Pogo, a professor in the departments of medicine and microbiology at Mount Sinai School of Medicine, in New York. Examining some four hundred to five hundred breast-cancer samples, she has found DNA sequences resembling MMTV that are not present in normal tissue or in other human cancers. She remains guarded about the implications.

Can You "Catch" Schizophrenia?

Microbes obviously can cause mental disorders — as syphilitic dementia, to name but one example, makes brutally clear. But most post-Freudian discussions of psychiatric dysfunction have tended not to invoke infection. Recently, however, some cases of childhood obsessive compulsive disorder (OCD) have hinted at a new set of possibilities. Children who have this disease may compulsively count the crayons in their book bags over and over again, or meticulously avoid each crack in the pavement, in order to ward off some imagined evil. Susan E. Swedo, of the National Institute of Mental Health, in Bethesda, Maryland, noticed strong resemblances between OCD and a disease called Sydenham's chorea, formerly known as Saint Vitus's dance, which, like rheumatic heart disease, is a rare complication of an untreated streptococcal infection. Streptococcal antibodies find their way into the brain and attack a region called the basal ganglia, causing characteristic clumsiness and arm-flapping movements along with obsessions, compulsions, senseless rituals, and *idées fixes*. Could some cases of childhood OCD be a milder version of this illness? The hunch paid off. In the early 1990s a new syndrome, known as PANDAS (pediatric autoimmune neuropsychiatric disorders associated with streptococcus), was recognized.

Some children with OCD get better when they are given intravenous immunoglobulin or undergo therapeutic plasma exchange to remove the antibodies from their blood. It is not known whether adult-onset OCD — whose most famous avatar was the germ-phobic Howard Hughes — also results from some sort of infection. But it is certainly provocative that a mental disorder can result from a lingering immune response. The phenomenon makes some people wonder about schizophrenia.

For years, amid the smorgasbord of theories about the etiology of schizophrenia, there has been recurring speculation about a schizophrenia virus. Karl Menninger wondered in the 1920s if schizophrenia might result from a flu infection. Later researchers pointed to data that showed seasonal and geographic patterns in the births of schizophrenics, suggestive of infection — though it must be said that the viral theorists were largely regarded as inhabiting the fringe. Genetic theories grabbed center stage, and by the 1990s most researchers were pinning their hopes on the genetic markers being identified in the Human Genome Project.

In Ewald and Cochran's view, evolutionary laws dictate that infection must be a factor in schizophrenia. "They announced they had the gene for schizophrenia, and then it turned out not to be true," Cochran said one day when I mentioned genetic markers. "I think they found and unfound the gene for depression about six times. Nobody's found a gene yet for any common mental illness. Maybe instead of the Human Genome Project we should have the Human Germ Project." Cochran is endorsing a suggestion made by several scientists in a recent issue of *Nature*. "I don't mean to say that the Human Genome Project isn't worthwhile for many reasons, but all the genes we've found have been for *rare* diseases. I don't think the common diseases are going to turn out that way."

Schizophrenia affects about 1 percent of the population and thus in Ewald and Cochran's scheme is too common for a genetic disease that profoundly impairs fitness. As noted, the background mutation rate — the rate at which a gene spontaneously mutates — is typically about 1 in 50,000 to 1 in 100,000. Not surprisingly, genetic diseases that are severely fitness-impairing (for example, achondroplastic dwarfism) tend to have roughly the same odds, depending on the gene. (In a few cases, however, the gene involved may be especially error-prone, resulting in a higher frequency of mishaps. One of the most common genetic diseases, Duchenne's muscular dystrophy, afflicts boys at a rate of 1 in 7,000, reflecting the fragility of an uncommonly long gene.)

From the fitness perspective, schizophrenia is a catastrophe. It is estimated that male schizophrenics have roughly half as many offspring as the general population has. Female schizophrenics have roughly 75 percent as many. Schizophrenia should therefore approach the level of a random mutation after many generations. (To explain this away, some genetic theorists have proposed that in hunter-gatherer cultures, schizophrenics were the tribal shamans

— desirable as sexual partners — and thus did not incur a reproductive disadvantage.)

No one has found a schizophrenia virus yet, but some think they
may be close. Following a tip from Ewald and Cochran, I typed
"Borna virus" into my on-line search engine and ended up with a
stack of scientific papers. Borna virus was first recognized as the
cause of a neurologic disease in horses and can infect nearly all
warm-blooded animals, from birds to primates. Horses and other
animals infected with Borna virus may exhibit depressed or apa-
thetic behavior, weakness of the legs, abnormal body postures, or a
staggering gait. Borna-infected laboratory rats exhibit learning
disorders, exaggerated startle responses, and hyperactivity, among
other things.

Royce Waltrip, an associate professor of psychiatry at the University of Mississippi with an expertise in virology, studies Borna virus.
Despite being leery of a rash of inconsistent studies associating
Borna virus with schizophrenia, Waltrip believes that "there is
something there, though I don't know if it's a perinatal infection or
an adult infection or what." When he started looking for antibodies to Borna in mental patients, he found that 14 percent of the
schizophrenic patients had antibodies to two or three Borna proteins, whereas none of the healthy controls did. Waltrip speculates
that Borna virus is not *the* cause of schizophrenia. "I think that
schizophrenia is an etiologically heterogeneous disease," he said. "I
think there are a finite number of ways the brain can respond to in-
jury. There are probably different routes to schizophrenia, and
there is probably more than one infectious pathway." One route,
he hypothesizes, is Borna virus.

Ewald and Cochran do not doubt that multiple pathogens
or multiple factors may be implicated in some broad disease syn-
dromes, among them schizophrenia. But they worry, in general, that the "multifactorial" argument has become too facile a response. "That's what they *always* say when they don't know the
cause of a disease," Cochran said on the phone. "They say it's
multifactorial. Ulcers and heart disease were supposed to be multifactorial. But they're infections! Tuberculosis was supposed to be
multifactorial. It's an infection!"

I happened to be visiting Ewald in his office when Cochran
called, so we were having a three-way conversation, with Cochran's

voice echoing over the static on a speaker phone. Outside the window the scene was shifting subtly into mid-autumn. Patches of orange and rust speckled the blue-green flanks of the Holyoke hills, and the students on the playing fields were wearing sweatpants.

But what about random accidents in utero as a cause of schizophrenia? I asked. Some kind of damage to the wiring?

"You'd have to say what caused the damage," Ewald responded, pointing out that the word *random* is often used to refer to something we haven't been able to understand. He noted once again how widespread schizophrenia is. "At this frequency — one percent of the population — we'd expect that natural selection would have led to protective mechanisms."

The same holds true for severe depression, Ewald believes. A tendency toward suicide doesn't make evolutionary sense in a world of organisms driven by the twin urgencies of survival and reproduction. The relentless engine of natural selection should have eliminated any genes that infringed on them. So why are these fitness-antagonistic traits still around?

This leads to a subject that Ewald is not shy about bringing up in discussions with colleagues and in professional lectures: homosexuality. Various pieces of evidence have been adduced in recent years, by prominent researchers, for some sort of genetic component to homosexuality. The question arises as to whether natural selection would sustain a homosexual trait in the gene pool for any length of time. The best estimates of the fitness cost of homosexuality hover around 80 percent: in other words, gay men (in modern times, at least) have only 20 percent as many offspring as heterosexuals have. Simple math shows how quickly an evolutionarily disadvantageous trait like this should dwindle, if it is a simple genetic phenomenon. The researchers Richard Pillard, at the Boston University School of Medicine, and Dean Hamer, at the National Cancer Institute, are not persuaded that natural selection would necessarily have eliminated a homosexual trait, and offer ingenious counterarguments. (And they note that historically the fitness cost may not have been very high, when gay men stayed in the closet, married, and had children.)

No one, of course, has ever isolated a bacterium or a virus responsible for sexual orientation, and speculations about the manner in which such an agent would be transmitted can be nothing more than that. But Ewald and Cochran contend that the severe "fitness hit" of homosexuality is a red flag that should not be ignored, and that an infectious process should at least be explored. "It's a very sensitive subject," Ewald admits, "and I don't want to be accused of gay-bashing. But I think the idea is viable. What scientists are supposed to do is evaluate an idea on the soundness of the logic and the testing of the predictions it can generate."

The Search for Telltale Signs

After I had spent time talking to Ewald and Cochran and reading back issues of the journal *Emerging Infectious Diseases,* everything began to look infectious to me. The catalogue of suspected chronic diseases caused by infection, according to David A. Relman, an assistant professor of medicine, microbiology, and immunology at Stanford University, now includes "sarcoidosis, various forms of inflammatory bowel disease, rheumatoid arthritis, systemic lupus erythematosus, Wegener's granulomatosis, diabetes mellitus, primary biliary cirrhosis, tropical sprue, and Kawasaki disease." Ewald and Cochran's list of likely suspects would include all of the above plus many forms of heart disease, arteriosclerosis, Alzheimer's disease, many if not most forms of cancer, multiple sclerosis, most major psychiatric diseases, Hashimoto's thyroiditis, cerebral palsy, polycystic ovary disease, and perhaps obesity and certain eating disorders. From an evolutionary perspective, Cochran says, anorexia is strikingly inimical to the survival principle. "I mean, *not to eat —* what would cause that?"

"In all these situations you look for little signs of infectious spread," Ewald said in his office. "Is there geographic variation? Temporal variation? Does it go up or down across decades? Multiple sclerosis seems pretty clearly infectious, because you have these island populations where there was no MS and then you see it spread like a wave through the population. And you have this latitudinal gradient . . ."

"Yes!" Cochran burst from the speaker phone. "The farther you get from the equator, the more common it is. It's three to four times more common if you grow up in Ontario than if you grow up in Mississippi. Some people have tried to say that's because Canadians are genetically different from Americans."

I downloaded a paper about extremely high rates of multiple sclerosis in the Shetland and Orkney Islands and other regions of Scotland, and I made a mental note of the many Canadian Web sites devoted to MS. Like other autoimmune diseases, MS looks suspiciously infectious for a number of reasons: epidemiological evidence of childhood exposure to disease agents, geographic clusters, abnormal immune responses to a variety of viruses, resemblances to animal models and human diseases with a relapsing-remitting course. And, in fact, a virus has been nominated: the human herpes virus 6, the agent of roseola infantum, a very mild disease of childhood. The connection, however, is by no means proved.

"No doubt everywhere people look there will be more and more examples of chronic diseases with infectious etiology," says Stephen S. Morse, an expert in infectious diseases at the Columbia University School of Public Health. "*Helicobacter* is probably the tip of the iceberg." Although we have wielded the tools of microbial cultivation for a hundred years, much of the microbial world is still as mysterious as an alien planet. "It has been estimated that only 0.4 percent of all extant bacterial species have been identified," David Relman has written. "Does this remarkable lack of knowledge pertain to the subset of microorganisms both capable of and accomplished in causing human disease?" Even the germs that inhabit our bodies — the so-called human commensal flora, such as the swarming populations of organisms that live in the spaces between our teeth — are largely unknown, he points out. Most of them are presumably benign, up to a point. There are disquieting suggestions in the literature of a link between bacteria in dental plaque and coronary disease.

"Some people think it's scary to have these time bombs in our bodies," Ewald says, "but it's also encouraging — because if it's a disease organism, then there's probably something we can do about it. The textbooks say, In 1900 most people died of infectious diseases, and today most people don't die of infectious disease; they die of cancer and heart disease and Alzheimer's and all these things. Well, in ten years I think the textbooks will have to be rewritten to say, Throughout history most people have died of infectious disease, and most people continue to die of infectious disease."

WENDY JOHNSON

Heavy Grace

FROM *Tricycle*

BOTH MY PARENTS DIED at the end of 1998, each of them on a Monday, a little less than three months apart. Although they had been divorced for forty years, they flared out together like two long-tailed meteors burning a nasty parallel gash in the cold dome of the winter sky.

Even though I have been practicing Zen meditation for twenty-eight years and working as a frontline hospice volunteer for ten, nothing helps. Nothing. The back of my head has been ripped off, and I'm immune to that unctuous snake-oil salve of "no coming, no going; no birth, no death" that well-intentioned Zen friends dab on my raw scalp. Give me good old Rujing from twelfth-century China any day, who, when setting fire to Elder Yi's funeral bier, cried out, "Ah, the swift flames in the wind flare up — all atoms in all worlds do not interchange."

I'm keeping to myself these days. Evaded by even the slightest desire to be civil, I am steadied by the unseasonable cold of the California winter, by rare snow, black ice, and saber-fanged wind. I take strange delight in the burned-to-a-crisp parts of tender plants frozen by the cold. I want all green shoots to die a long, slow death. At the same time, I spend my Mondays planting dormant, bare-root plants for my parents.

My mother died in our family home in the Adirondack Mountains of upstate New York. My sisters and I, our mates and our children, planted a Prairie Fire crab apple for Mom just outside her bedroom window a week after her death. It was a strong tree with burnished mahogany bark and laden with tiny, blood-red crab apples.

At home in California I kept planting for Mom, every Monday for seven weeks, this time around my house in a patch of abandoned, shady ground. It's OK, I reasoned to myself, my mother is a shade, so I cleared the neglected land and wove a tapestry of grief out of deep-shade plants: Corsican hellebores, perfumed daphnes, and wild ginger interlaced with dark violets from Labrador. "Newfoundland," I said under my breath, and kept on digging.

It's harder to plant for my dad. His death is fresh, ragged, vicious. I am planting seedling redwoods for my father, trees started from seed we collected a few years ago when he was still strong and coming to Green Gulch every winter to plant forest trees with us in February. Cultivated gardens made my father bolt; he longed for the clean sweep of the Fire Island barrier dunes or the mangy, unkempt silhouette of ancient forests jutting up through winter fog. So I'm planting redwoods for Dad, far beyond the garden gate. Out here it is premature to think of grace, even heavy grace. The ground has just begun to thaw. Give me my spade and let me dig.

KEN LAMBERTON

The Wisdom of Toads

FROM *Puerto del Sol*

I WAIT IN LINE for my weekly shopping at the inmate commissary
and watch a man with a toad in his hands. We had rain this morning, a tremendous predawn thunderstorm of the kind that never
fails to keep the prison yard on lockdown. Wind and rain blurred
the perimeter lights, while lightning threatened to shut down the
power system altogether. It was a storm for toads. Later, after the
storm passed and we were allowed to begin our day, I crossed the
yard, navigating chocolate pools and streamlets to pick up my bag
of groceries. Today: coffee, a loaf of wheat bread, a jar of crunchy
peanut butter, plastic hangers, and some writing materials.

The inmate had found the toad where it breached the flooded
Bermuda in a low area between the store and the classrooms. Now,
he cradles it as he would a letter from a wife, or a photo of his baby
girl. The two seem to communicate a mutual fascination. One cap-
tive to another. It's a benevolent scene, and an unlikely one on this
gray day wet with humidity and sweat. At the head of the line a
guard notices the scene too, but his face is contorted. It is a hard
look, disguised of humanity. *This is prison,* it says. In mock concern
for the animal he shouts, "Put the frog down! It's done nothing to
you!" Oh, but it has, I think. It's charmed him, here in the middle
of this desperate place. And maybe for a moment, he's escaped
into the memory of something — a childhood pet and playmate,
perhaps. I guess this from the way, even now, he watches the toad
back in the grass. And the look of his vague but plain smile.

I dig for a toad in the visitation sandbox, to the expectation and delight of my daughters. The amphibian is huge — a Sonoran desert

toad — and wedged tightly under the concrete sidewalk. I stick my hand into the dark and clamp down on its head. It has no neck. I work my fingers like forceps laboring against a stubborn birth. The toad puffs itself up and digs in. It won't come easily. I am nervous about it. The guards wrote up an inmate last week for doing what I'm doing. They said he was stashing drugs; he insisted it wasn't the case, that his kids said something was moving down in the sandy hole. Disciplinary court couldn't find him guilty without any evidence, but he still spent several days in the "hole" on investigative lockdown. Solitary confinement. It's like going to prison in a prison.

It takes both hands to hold the toad. It bucks and squirms and finally empties its bladder over my fingers, and my girls jump back and scream. "It's just water," I tell them. "It's normal. It wants me to let it go." I put the toad down in the sandbox and the girls encircle it. Karen watches for the guards, concerned they might disapprove. Animals can be unpredictable and the guards insist on control. The toads are deadly, too, so they inform us every time one appears and they notice our interest.

The toad lifts itself on stubby legs and moves forward one body length. This is not what I would consider a hop, more like a push. The girls back away but keep an interested perimeter. It pushes again. It's all warts and eyes, the latter like clear golden beads attached to an ugly green bag of liquid. Melissa, blond, blue-eyed, and my youngest, can't handle it anymore. She climbs into the arms of her mother and complains, "It's staring at me. It's staring at me."

It's possible that the inmate the guards wrote up *was* stashing drugs. I know people on this yard who lick toads for the intoxicating effects. As a defense, Sonoran desert toads secrete a toxic milk from large glands behind each eye. The poison can cause temporary but violent fits and paralysis in coyotes and dogs when the animals bite or chew on the toads. Unless you stick a finger in an eye after handling one, the toads aren't harmful to people. And licking the slime-laden warts will only make you sick, not high. Those who know what they're doing don't lick toads; they smoke them. Burning the poison and then inhaling it can send the smoker on an intense hallucinogenic mind-warp not unlike that produced by LSD. Toad-smoking aficionados, a growing subculture even outside of prison, don't actually harm the toads. They squeeze the glands to collect the poison, then dry it and mix it with tobacco (toad rollies) or burn it in a pipe.

I've never smoked a toad before, but I've tasted the toxin. It's as tongue-numbing as Novocain. If inhaling the smoke is as mind-numbing, I can understand why the toads attract some prisoners. They are a way out of this place. They are ugly green bags of chemical-induced freedom. Toads, however, offer more than drugs as a way of relief from walls and bars.

On warm evenings Sonoran desert toads the size of grapefruits wait for darkness in the shrubbery outside our cells. Lights attract insects, which attract toads. Occasionally men will gather around with guitars, and they'll sing lonely ballads about unfaithful wives or distant girlfriends. Mexican music prevails, which seems most appropriate for this place. As the toads appear, the men catch beetles to feed them. When they run out of beetles, they'll toss crickets, scorpions, ripe olives, even juvenile toads. The cannibals snap up everything that moves, including lighted cigarette butts. This offering the toads reject, backing away, drawing in their eyes, scraping their tongues with tiny hands. A cruel form of amusement but not the worst. I've seen the blackened, twisted bodies of giant toads dangling lifeless from curls of razor wire, as if a shrike had made its larder there.

The toad in the visitation sandbox patiently humors my girls as they construct ponds, walls, and castles for it to live in. Then, finally, it escapes and dives into its underground chamber. I can't blame it. As comfortable and protective as enclosures may be, prison still makes me want to burrow underground like a toad. We leave it alone for now.

I came to prison when Jessica was four years old, Kasondra two, and Melissa only a few months. Melissa took her first steps in prison visitation. I knew I would be at most just a weekend presence in their lives, hardly a father except by title. I wanted to influence my children as much as possible. They were already too much like their mother: blond, lithe, intelligent . . . girls. Not that this is a bad thing, of course. Still, they carry half my genes, even if it doesn't show. They would need me, I reasoned, to balance their dolls with balls, their ribbons and bows with insects and toads.

From the beginning, I took them on missions for toads in the visitation park, a large enclosure of trees, grass, and picnic tables attached to the main visitation area that now stands idle, unused, although it's kept neatly landscaped. We searched for them in the deep grass of a drainage ditch, beneath oil drums used as garbage cans, and along a dirt strip at the perimeter fence. Most were small Sonoran desert toads and Great Plains toads. Occasionally, we'd uncover a spadefoot. My daughters would fill the pockets of their Osh-Kosh overalls with the cool, thumb-sized nuggets of activity and laugh as the toads chirped rapid-fire complaints from under their clothing. They had bold intentions of smuggling them home. Little criminals. Typically, however, they would rebury their bug-eyed jewels in the sandbox and then look for them from week to week and, often as not, locate them again. Digging up toads in visitation was our game. Other fathers played checkers or Scrabble with their kids; we went hunting for buried treasure.

Nature writer Christine Colasurdo says we live on one layer of the earth at a time. It just happens that my one layer has razor wire horizons. Despite this, I still wanted my love for wilderness to seep into my children. Prison wouldn't last forever, and I looked forward to a time when we could all go camping and hiking and fishing, things I missed sorely and hoped to share with them. I wanted them to accept the whole experience of wilderness, too, not the window-seat tour but the *adventure,* the dirt, sweat, and hunger. I would bait their hooks as long as the girls would deal with mosquito bites and cold, wet feet.

Until then, I could prepare them with whatever bits of grace prison sent us, bits like toads. For now, the toads in visitation would be our way to wilderness, a squirming, hand-held means for me to connect my children to nature. I couldn't be sure where it would lead. But as toads moved into the girls' bedroom and became pets, science projects, and essays, I felt less concern about balance in their lives.

Then came my miraculous (if temporary) release and that one summer my daughters and I found the spadefoot toads. The previous December, after serving nearly eight years of a twelve-year sentence, a superior court judge overturned my conviction and sent me home to my family. The decision didn't come easy, not for the judge and not for my wife and lawyers. After I went to prison, Karen, while caring for three small children, went back to school to study criminal law. She specialized in legislative histories, focusing on the law behind my crime. She researched the similar statutes of other states and met with legislators. She spoke at conferences, appeared on radio talk shows locally and television talk shows nationally. She began to use the law and the media. When my arresting officer approached her, disturbed by the length of my sentence and offering to testify on my behalf, Karen sought out a law firm interested in our case. The Sherick law firm liked Karen's work and agreed to both hire her as a paralegal and take me as a client. What's most amazing to me still is Karen's response to my crime. She could have divorced me, a common reaction to behavior like mine. I had humiliated her publicly, betrayed her in the worst way a man can betray a woman, a husband a wife. I deserted her for someone younger. And it was worse than this, criminal. I was a teacher; she was my student. What grew out of mutual affection led to my obsession. I thought that running away with her would solve everything. We could start a new life together. But the romance ended with my arrest two weeks after we left Arizona. She was only fourteen years old.

The judge took three months to make his decision, basing it on two days of testimony at my hearing for postconviction relief. Evidence the state had collected to originally convict me, together with new testimony, including that of the arresting officer, my lawyers now used to win my release. At my resentencing, the judge said it was time for all of us to move on with our lives. He placed me on probation for the remainder of my sentence and allowed me to go home. Unfortunately, the Arizona Court of Appeals would not agree with him. Eighteen months after my prosecutor appealed the decision, three justices ordered my return to prison. The appellate court does not go easy on sex offenders.

The first summer of my temporary release the monsoon season began with a terrific thunderstorm, which darkened most of Tucson late one afternoon. The wind came first, raising sheets of dust and sticking leaves and twigs to windowpanes, which rattled and vibrated as if to loosen the debris. As the wind finished its choking assault, it charged in with rain, stinging rain, huge, pelting, steel-ball-bearing rain, rain that cratered the ground and tore it away in muddy streams. For most of the year the arroyo in our backyard is only sand — bright, eye-squinting quartz grit. But on this afternoon the wash raged with mud and foam and uprooted vegetation carried by the afterbirth of the storm.

By evening, the water had receded and the spadefoot toads began calling out in the poststorm quiet. Theirs were the cries of motherless lambs, urgent, as if they understood the dictates of shrinking pools in the desert.

My daughters were nervous about leaving the house after the storm, particularly after dark. But the toads drew them, too. Something mysterious was happening down at the arroyo; we could hear it through the drizzle. I held on to a flashlight and dip net while the girls clung to my arms, and we worked our way down the road and up the wash. Shrubs and grasses, bent and torn, betrayed the recent passage of the floodwater, as if some enormous snake had just slid past us. But the wash was now only a broken chain of pools. We found the spadefoots in a truck-sized hole gouged out of the wash by the storm runoff and then filled with muddy water. I slipped into the pool, sinking waist-deep, and netted the first pair of toads I saw locked in amplexus. We quickly returned to the girls' aquarium and released the mating toads. In the morning we freed them, but not before they left us with a mass of fertilized eggs, hundreds of tiny black beads suspended in clear jelly.

This was the summer we discovered that spadefoot toads are one of the desert's miracles. For years they rest underground, barely alive, escaping a hot, dry landscape that would otherwise reduce them to wood chips. We learned that in this state they eat nothing. They drink nothing, storing a third of their body weight in water in canteenlike bladders. For most of their lives, the amphibians are mere seeds. Then, awakened by the deep, pounding rhythm of a summer thunderstorm, they emerge en masse to breed. It's as if the right mixture of water and sand produces toads by spontaneous generation.

The eggs in my daughters' aquarium began hatching in less than thirty-six hours, each tadpole spinning and kicking within a clear capsule, fighting to free itself from the jelly. We estimated that we had more than eight hundred tadpoles, tiny wrigglers all head, gills, and tail. In the arroyo, the pool was sinking into the sand, marooning globs of eggs on its banks like spilled caviar drying in the sun. But many of the eggs had hatched, and thousands of black commas punctuated the tea-stained water.

We had read that spadefoot toads are record holders for metamorphosis. Normally, it takes them about two to six weeks to change from hatchling to toadlet, depending on things such as water temperature and available food. But we also knew that the tadpoles could adjust their growth according to the duration of their pond when it isn't too crowded. If the alternative is getting baked into the mud curls at the bottom of a puddle, the tadpoles can sprout limbs and crawl onto land in less than eight days. Sometimes, we learned, it might be the only way to survive an unpredictable desert pond.

With this information, the girls decided to keep a notebook to record the tadpoles' metamorphosis, comparing those tadpoles in their aquarium with those in the arroyo and logging the differences in temperature. Our adventure was becoming a science project. Their notes were interesting, if not entirely conclusive. In less than a day the girls' tadpoles had doubled in size. In five days they were as big as peas; after six they had hind-leg buds. At ten days the tadpoles' round bodies had deflated into lumpy toad forms, front and hind limbs flexing new muscles as their tails dwindled. As they grew from swimmers to floaters, my daughters scooped them up and placed them into a terrarium.

The pond in the arroyo had been shrinking steadily since the storm, the tadpoles there becoming frantic with motion as the temperature increased and the borders of their world descended on them. At eight days the pool was a roiling vat of tadpole soup. On the ninth day it was gone. The girls wrote that when they approached the place, dozens of small toads hopped across the wet sand, dragging limp tails behind them.

It was interesting for us to notice that the spadefoot frenzy in our wash coincided with the emergence of a congress of insects. Most significantly, the thunderstorms had triggered termites. On the same evening when the amphibians had pushed to the surface to begin their mating and egg-laying cycle, hordes of winged termites boiled out of their nests for their aerial nuptials. Termites are small packages rich in fat, enough calories in one or two nights' gorging to ready the toads for their long sleep and instantaneous fertility.

Melissa discovered that not only were termites easy to find and collect but that baby toads relish them. The toads targeted on the white, moving bodies, lapping them up with tiny, sticky tongues. As the toadlets quickly grew, the girls fed them larger and larger fare — ants, sow bugs, beetle grubs. By the end of summer they had the toads accustomed to gulping down tangled knots of pet store blackworms served on a toothpick.

It's late October now, less than three months since my return to prison. A Pacific cold front is delivering its burden of cold air and rain and wind, sticking the blackbirds to the ground and muddying the track. I haven't seen any swallows, those deciduous birds, in a few days. I suspect they've left on the leading edge of the storm, and already I miss the way they dip and cut across the Bermuda on razor-blade wings. Recently, we've had record highs in the eighties, and the swallows have been massing. I counted more than a hundred flying circles over the field, the most I'd ever seen here.

Today, a rare sight. A dozen or so ducks fly a perfect, south-pointing V against a concrete sky — a typical fall scene that seems misplaced in this desert.

Another rarity. Off to the side of the exercise track I find a toad. It's perfectly still when I pick it up, an ice cube in my hand. I recognize the squat, goblinlike amphibian immediately: bright yellow-green skin and swollen eyes. A Couch's spadefoot. It's alive; the eyes shine and I can see deep inside them. I carefully turn it over and look for its digging tools, the namesake black half-moon "spades" on both hind feet. Strange, I think, this cold-weather toad. It should be deep underground. It's been a while since I've noticed any signs of them at the margin of the track, the pushed-up mounds of dirt, like little impact craters, where spadefoots sink backwards into the ground in slow spirals. This one must have been confused by the storm, the pounding rain, or perhaps the tremors of a passing delivery truck, or a combination of both. I've heard that the deep vibrations of off-road vehicles can wake the toads out of season.

I spend the evening with the toad, thinking of my children and trying to sketch it. It won't cooperate. I can't tie it down as I've done with lizards, and the glue I've used on the feet of ants and beetles is out of the question. With my notebook in my lap and a pencil in one hand, I hold the creature between a thumb and three fingers as it warms and squirms under my desk lamp. Now it can see deep into my eyes. When its squirms turn into hiccups in staccato, I put down my pencil. The toad sings, filling my cell with the same passion expressed in those lonely Spanish ballads. It is the wisdom of more than toads that companionship begins with song. But desire sometimes goes unheard and unanswered out of season. Or out of place.

I try not to think about this. Instead, I hold the toad and allow myself to listen to the wildness, the freedom in its voice. It's not an escape, more a kind of passage. It guides me to that nighttime foray with my girls where their small hands held those toads of summer.

PETER MATTHIESSEN

The Island at the End
of the Earth

FROM *Audubon*

THE SHIP SAILS from Ushuaia, Argentina, at 6 P.M., due east
down the Beagle Channel. To the north and south, the mountains
of Tierra del Fuego are dark, forested, forbidding, showing no
light or other sign of habitation. Already, a soft swell tries the bow,
and a gray-headed albatross planes away across the rolling wake.
Next, a black-browed albatross appears out of the east, where high
dark coasts open on the ocean horizon and the last sun ray glints
on the windy seas of the Drake Passage.

Across the strait lies Staten Land, a seagirt mountain ridge so
steep and rugged that it would be difficult to find a place to land.
Staten Land is land's end in South America, where the mountains
slide beneath the sea. They will surface again 1,000 miles eastward,
in the farthest region of the South Atlantic, as the icy peaks of
South Georgia Island.

Night falls as the dark land disappears. In the wash of sea along
the hull, the ship's diesels are strangely quiet, with scarcely a vibration. Toward midnight come the first hard bangs and shudders of
the ship's adjustment to the open ocean, and when daybreak arrives at 5:30 A.M., she is out of sight of land in the toiling iron seas
of the Cape Horn Current.

The *Akademik Ioffe* is a 384-foot research vessel with a crew of
fifty-two. Built in Finland in 1989, she has been chartered to Victor
Emanuel Nature Tours, the wildlife-safari company out of Austin,
Texas, that organized this out-of-the-way voyage to South Georgia

and Antarctica. The ship has a good small library and a sauna, but most of the passengers spend the day looking out to sea.

The coastal birds have now fallen away, and the powerful pelagic birds have found the ship, crisscrossing the waves on long and pointed wings as stiff as boomerangs. These ocean fliers, from the great albatross to the diminutive storm petrels, belong to the great order Procellariiformes (from the Latin *procella*, for "storm" or "gale"). All members of this large and varied tribe share external nostrils in the form of tubal structures on the upper mandible, which extract the salt from ocean water, a remarkable evolutionary adaptation that permits them to spend most of their lives at sea. Young wandering albatross may cruise the southern oceans for five years before returning to the breeding grounds, without once setting a webbed foot on land.

The wandering albatross, *Diomedea exulans,* is the greatest of its clan, with the broadest wingspan (eleven feet) of any bird on earth. Arching down the sky to vanish behind a wave, curving high again like a white cross, it excites cries of wonder. "I now belong to a higher cult of mortals, for I have seen the Albatross" — surely, *exulans* is the species celebrated by Robert Cushman Murphy, whose two-volume classic, *Oceanic Birds of South America,* was published in 1936. A pioneer work on these little-known ocean wanderers, it filled me with a longing to journey to South Georgia, describing the astonishing bounty of life on the shores of those white, icy peaks lost in midocean, where one might have thought no life could exist at all.

South Georgia. First sighted in 1675 by a London merchantman blown far off course at the farthest curve of the undersea ridge called the Scotia Arc, it was not reported a second time until nearly a century later, by the Spanish ship *Leon,* in 1756. In England, Captain James Cook would give no credence to the first report and very little to the second, at least publicly, but eventually he felt obliged to credit the Spanish captain Gregorio Jerez with inspiring the island's true discovery two decades later by a stouthearted Englishman, namely himself.

Though Cook would perish in Hawaii in 1779, his crew brought word back to England of the great companies of seals and whales around South Georgia, where the first sealers arrived in 1790. In

1800, 122,000 fur seals were slaughtered there. Twenty-three years later, the American sealer Captain James Weddell wrote of the fur seals that "these animals are now almost extinct." When the fur seals were exhausted, the sealers turned their clubs to the huge elephant seals, whose blubber was boiled down for its oil.

Today South Georgia has one of the greatest concentrations of seabirds and mammals anywhere on earth because of the astonishing abundance of the small marine organisms known collectively as plankton, on which virtually all species of petrels, penguins, whales, and seals depend. Why South Georgia is extraordinary in this respect is best described by the oceanographer Sir Alistair Hardy, of Robert Falcon Scott's 1900–1904 *Discovery* research expedition, which preceded his doomed trek to the South Pole:

> South Georgia is a long narrow island, some 100 miles long by some 15 across; it is placed almost at right angles to the main westerly drift coming up the Drake Straits. . . . Where the main ocean current from the west strikes the continental shelf of South Georgia [is] an upwelling of water rich in phosphate from the deeper layers on the west side of the island, and it is here that we get the densest growth of diatoms, which are carried round either end into the area behind the island. . . . Here in this sheltered water are all the Euphasians, young and old, which feed on the diatoms. . . . This is, perhaps . . . why South Georgia, so peculiarly situated, should be one of the richest whale-feeding grounds of the world.

As late as the 1930s, whales were still abundant all around South Georgia, where the industry was supported by rorqual, or baleen, whales — notably the blue whale, the humpback whale, and the finback, or fin whale. Only much later, when these larger species became scarce, did the whalers start hunting smaller species such as the minke. Today the Antarctic Ocean is an international whale sanctuary, now that all its large whales are gone.

On January 24, 1998, in a bright dawn more than two centuries after Captain Cook's departure, the snow peaks and glaciers of South Georgia glisten under a clear blue sky as the ship moves southeast along the coast past Stromness Bay. In the first full sunshine of the voyage, the mountainsides are patched with verdant greens. Black cliffs fall steeply to the water at the mouth of the great Cumberland

Bay, and dead ahead, closing the eight-mile length of the blue bay, glittering ice walls form the leading edge of the Nordenskold Glacier, which soars away through thin and shifting mists to Mount Nordenskold, 7,725 feet above sea level. Like the Malvinas, these are true mountains, not volcanoes. The island's highest peak, Mount Paget, rises out of the sea from the ocean floor thousands of feet down to a white peak almost 10,000 feet above the ocean surface.

Glaciers descending between towers of black rock rule this mighty and ferocious landscape, including tidewater glaciers that come right to the water and "calve" small, irregular chunks of ice into the sea. All alone here, a half mile off the cliffs, the white *Ioffe* seems strangely diminished, even humbled. Between ship and shore, the sea is a clear jade green from the sheer density of organic matter. At this time of year, the seawater is supersaturated with oxygen, as much as 110 percent, a condition that produces high densities of tiny diatoms as well as high concentrations of hydrogen and iron, phosphates and nutrient salts, which in turn support the clouds of krill.

It is austral summer, and the coastal cliffs are broken by the grassy headlands, slopes, and benches used by the nesting seabirds. Below the cliffs are black-rock beaches, and here the white breasts of king penguins shine against the stones. Nearer, the sun catches the gold ear patches behind the eye of swimming members of this splendid species, as well as the yellow head tufts of the much smaller macaroni penguins, which surface here and there among the dark, round, shining heads of the Georgian fur seals. Antarctic prions go twisting past in their small scattered companies, like blown confetti, and overhead fly kelp gulls and Antarctic terns — the first coastal species seen since the ship left Tierra del Fuego.

In King Edward Cove, under a snow peak, lies the rusty-roofed hulk of Grytviken, established in 1901 by the Norwegians as the first whaling station in the Antarctic. With a labor force of perhaps three hundred men, the factory winched sixty-foot whales — as many as twenty-five every day — onto its slippery platforms to be flensed and cooked, not only for the whale oil in the blubber but also for bone and meat meal, and meat extract. Other Norwegian whaling stations thrived at Stromness until 1931, when the last South Georgia factories closed down for want of whales.

Today the station grounds have been reoccupied by the king penguins, which have joined a few introduced reindeer in their sad wanderings through the old ruins. Huge elephant seals lie in groaning rows by the shed walls, and a lively nursery of fur seal pups rush about on the grassy slope, which rises toward the cemetery where the great Antarctic hero Ernest Shackleton is buried.

Sailing from Grytviken on a new South Pole expedition in November 1913, the *Endurance* was beset by the pack ice in the Weddell Sea on January 19, 1914. Later that year she was crushed by the shifting pack ice after drifting northwest about 1,500 miles. In early November, Shackleton attempted a trek of 350 miles to Paulet Island, where an earlier expedition had left a hut. When this failed because of treacherous ice, the party returned to the ship, which sank on November 21. The men took to the ship's lifeboats, which they dragged and rowed hundreds of miles before reaching Elephant Island, named for the great lugubrious seals that would help keep the crew alive. Camp was made in a cave near a colony of gentoo penguins, and two days later, on April 24, 1915, leaving the remainder of the crew to pray for them, Shackleton and four companions set out in the remaining lifeboat for South Georgia.

Arriving there half dead from exposure, they were wrecked in the surf on the rocky and desolate windward coast. Soaked through, frozen, and exhausted, they climbed thousands of feet up the island's ice-bound central range and trekked southeast among the peaks for eighteen miles, making a dangerous descent of rock and ice to reach the Stromness whaling station, where the *Endurance* had been given up for lost. Scarcely recovered, Shackleton took sail as rapidly as possible for Argentina, from which he made two desperate attempts to reach his stranded crew. He accomplished this in a third effort on August 30, 1916, bringing his crew back to Britain without losing a man.

Shackleton died in Argentina in 1922, and his wife returned his body to South Georgia. At Grytviken, from which the *Endurance* had sailed in 1913, we drink champagne toasts at the hero's graveside to a chorus of deep groans and blarts from the lounging elephants.

In 1901 the first whalers at Grytviken reported that so many whale spouts could be seen from land that the hunters had no need to leave the bay — 174,000 whales are said to have been slaughtered out of that station alone. Last year, a young member of our ship's staff trekked for a month across these mountains, keeping an eye toward the sea looking for whales, and never saw one. Tim and Pauline Carr, a British couple who tarried here to help establish a small museum in the course of a round-the-world adventure on a small sailboat, tell us they have spotted only two whales from the shore in the past five years.

The Carrs are the first family to inhabit South Georgia since the whaling days. Their wooden vessel, a twenty-eight-foot gaff-rigged cutter without auxiliary engine or radio, celebrated its centennial last year. Still awaiting their departure, the pretty *Curlew* is tied alongside the half-sunken hulk of the old whale catcher *Petrel,* but the Carrs, who love this beautiful, austere place, are looking for a means that might let them stay. In the old days, South Georgia was enclosed by sea ice during winter, but for some years now no ice has formed, presumably because of global warming, so the small community they hope to found could be supplied by ship.

South Georgia was the first place "retaken" by that old warhorse Mrs. Thatcher in the cynical and absurd Falklands War between Argentina and Great Britain — a less than bitterly fought campaign, since South Georgia at that time was uninhabited. A British presence was established by a postmaster and a small garrison of Gurkha soldiers, who live in new red-roofed white barracks, west of the drab, cold, silent factory, and whose only afflictions in this hostile landscape are homesickness and boredom. In a rare engagement with the enemy — the first and last one in its history, for all I know — the garrison (which represents "the South Georgia government," a ghostly body seated in the Falklands/Malvinas) officially forbids our young Argentine stewardess and her brother from setting foot upon this lonely shore.

That afternoon, duly registered with the South Georgia government, the *Ioffe* sails out past Barff Point and returns up the coast. At Fortuna Bay, carved out over long millennia by the Fortuna Glacier, the east wall of the bowl is brown with shale, but on the west slope, beneath an ice field dirtied by blown dust, soft green moss and high tussock grass have taken hold, and on a rise is a colony of gentoos. In a squadron of four trusty inflatables run by the expert boatmen, we are set on shore.

Reconnoitering, I make my way uphill through the sprawled elephant seals. Despite their vast, jiggling avoirdupois, these huge creatures have hauled themselves high up into the tussock grass, two hundred yards above the beach. This is a commendable feat for an animal that can reach a length of twenty feet and weigh two tons and must hump and haul itself on its belly using fore flippers alone, since the hind flippers serve only as a rudder, useless on shore. In the sea, this giant among southern seals is powerful and agile, diving as deep as 4,000 feet in pursuit of squid and fish. The elephants open their jaws with weary sighs and cavernous yawns, peering through strange, dark, lustrous eyes — blind-looking because no pupil is visible (and eerily reminiscent of the black moon eyes of the great white shark). Despite an equable nature, the elephants utter prodigious roars as well as barks and snorts to dismay intruders, but they are not aggressive unless a human gets too close; then they may rear up in protest.

Far away up a delta of small streams that descend from the glacier, a broad area of white turns out to be the snowy breasts of a large flock of king penguins — some 7,000 pairs, by a rough count. I walk down to the beach and make my way through feisty fur seal pups and strolling companies of kings, then walk up the gravel stream a mile or more to the great central flock, swooped on by the screeching terns most of the way.

On this rare day of sun — the sixth sunny day in the past three months, as the Carrs informed us — most of the birds stand in the icy stream, and some have trekked all the way up onto the snows. The panting penguins are overheating and stand with their backs to the wind, lifting smooth flippers and even their small tight feathers to increase ventilation. (In severe cold they tuck everything in, feet and bill, facing into the wind to hold the feathers closed, since feather oil provides most of their insulation.) Many of them are in molt, including the big fluffy brown chicks. These stand disconsolate, peeping and chirping in their sweet, rich voices; the parents distinguish them by voice, not by appearance. They trudge along after the adults, bills pointed down, eyes to the gravel, in a manner that says, "Well, *this* isn't much fun!" So fluffy are they that their short tail is scarcely visible; they look as if they cannot quite lower their wings. Even the chicks that are mostly finished molting have tufts of brown down on the breast or back, and one is clean and beautiful except for a last tuft on one side of the head, hopelessly foolish and appealing. As yet, they neither dive nor swim well, and in the salt water they may be preyed on by skuas and giant petrels, which harass and peck them until they are exhausted.

After a few days of rough weather and fierce williwaws, the morning of January 27 dawns with whales and sun. Three humpback whales blow off the starboard bow, and they do not sound as the *Ioffe* comes abreast but porpoise easily along in no alarm or hurry. When at last they sound, their great gleaming black flukes lined white beneath, they rise slowly and majestically against the mountains all around, curving in high, graceful arcs and sliding silently without a splash into the sea.

Swirling snow mist shrouds Mount Paget as the ship edges her way into and out of Stromness Harbor. There, the three old whaling stations lie decrepit, like small, remote military outposts left behind by war. We gaze awestruck at the long snow glacier down which Ernest Shackleton and his frozen, ragged band had staggered after crossing the ice peaks from the windward coast where their small boat had been wrecked upon their return.

On this brilliant morning, the first snow petrels are sighted, two pure-white vagrants from the southern continent, and the beautiful pintado petrels, painted in motley browns and whites, and the local breed of the blue-eyed cormorant. At noon the ship turns inshore past black-rock reefs into Gold Harbor, formed by two glaciers whose white ice cliffs are hundreds of feet high; the glaciers rise and rise and rise to frosty horizons far away between black peaks. Along the edges of the bay, at the foot of steep, bare slopes of scree and grasses, are the shining browns of elephant seals and sea lions. In the shallows, oblivious of bathing penguins, the young fur seals play in galloping splashes and great flurries, chopping the surface white like feeding fish.

On shore, I negotiate an uphill path through a large herd of elephants, forty or more. They lift their wide jaws in great gasps, staring sightlessly through those strange moon eyes. On the gravel beach, in the cooling shallows of a glacier stream, fourteen are stacked side by side like fraternal two-ton loaves. For a mile upstream toward the glacier, the gravel bars are swarming with king penguins, an immense colony solid-packed at the far end of the beach and inland up the streams — in fact, uncountable, though we all agree on a minimum of 50,000 birds in sight at once. Many are still feeding their brown chicks, which are now a year old and still molting. With the king penguins are hundreds of gentoos, terns, gulls, ducks, and blue-eyed cormorants, with light-mantled sooty albatross gliding overhead in pairs. One scarcely hears them over the constant yawp and rumble of the mammals that are scattered the whole distance of the long beach and far back into the tussock grass behind.

The fur seal congregation here includes a group of bull-necked adult males, or "beachmasters," growling and charging. At times one must growl back, holding one's ground and, if necessary, jumping with arms spread, as our ancestral hominids are thought to have done to protect themselves against larger, faster, and better-armed animals. Such behavior will usually halt the charge, but one must be ready to leap sideways, and also to avoid a sneak attack once it has passed. One sees the dog relative in these animals — a dog of perhaps three hundred pounds.

The mammals rule for a considerable distance up the steep side of the headland, which we climb in search of the scattered ledge nests of the light-mantled sooty albatross. Eventually two are located and photographed, and also nests of skuas and the snowy sheathbill, which seeks nest sites beneath overhangs of the cliffs. A bird of the littoral, or foreshore, the sheathbill is one of only two species in its genus and family, occurring all the way around the world in austral latitudes. In size and general aspect, and even in its flight, it resembles a heavy white domestic pigeon, but at close hand one sees the bare skin on the face that is characteristic of many carrion feeders (which would otherwise have difficulty keeping their head feathers clean). Everywhere, this species acts as a nest robber and a scavenger, preying upon the eggs and chicks of penguins and other birds.

Each little while, the din and cries of the marine mammals and the tidal whisper on the gravel of the beach are broken by the loud crack and thunder of the calving glacier, like dynamite resounding in a rock quarry, causing frozen dust to rise out of the icefall after each explosion. With the amphitheater of black peaks and the dark-blue wind-chopped glacier bay opening into the circumpolar seas, one is awed and astonished not only by the sheer biomass but by the urgency of life renewal in this habitat. In late afternoon, as we depart, I look back a bit wistfully at Gold Harbor, knowing how unlikely it is that I shall return in this life to this remote and magnificent island.

Before departing South Georgia for Antarctica, the *Ioffe* enters the Drygalski Fjord, a narrow aperture that penetrates some eight miles from the open sea into the island's mountain heart, all the way to the great glacier under Mount Macklin. There are few birds in the narrow fjord, but far inland snow petrels come and go like lost white spirits between high, dark walls. Offshore, large flocks of white-chinned petrels cluster and dip on the rich plankton, and a few wandering albatross carve the twilight. In the growing darkness, I climb to the bridge to see Cape Disappointment at the east end of this mighty island, ringed by explosions of white surf. There is no beacon, nor any sign of man.

South latitude 54 degrees, 42 minutes. South Georgia, one of the most remote places on earth, is the last outpost in a great emptiness of ocean. Were this ship to proceed due east, circumnavigating the austral oceans of the planet, she would pass well south of the Cape of Good Hope, Tasmania, and New Zealand before making her first continental landfall near Cape Horn.

The ship rounds the rock islets that lie off Cape Disappointment, and the ghostly snows of South Georgia's windward coast come into view. The mountains fade in the starboard mists as the ship bears southwest for Elephant Island, off the Antarctic Peninsula. The bow rises and falls on the long swells of the Scotia Sea, smote by the night wind and the stygian blackness of a fast-moving squall, crossing drowned mountains.

CULLEN MURPHY

Lulu, Queen of the Camels

FROM *The Atlantic Monthly*

JULIAN SKIDMORE IS lithe and petite, with small wrists and delicate features, and a serene but determined countenance. Watching Skidmore at work for a while, her auburn hair held back by a blue ribbon, a glint of light catching the small pearl in each earlobe, I was reminded of Gainsborough's portrait of the young Georgiana, Duchess of Devonshire. Then Skidmore removed her left arm from a camel's rectum, peeled off a shoulder-length Krause Super-Sensitive disposable examination glove, and said, "Can I make you a cup of coffee?" She had completed eight of the morning's sixteen ultrasound scans. It was time for a break.

Skidmore, an Englishwoman known to everyone as Lulu, has emerged during the past few years as among the foremost practitioners in one of the world's more improbable growth industries. There are many reasons why *Camelus dromedarius,* the single-humped dromedary camel of Africa, Arabia, and southern Asia, might have deserved to become a focus of scientific investment. To begin with, about 14 million of these animals roam the planet. The dromedary camel is a baroque masterpiece of biological engineering. It is relied upon by millions of people for meat and milk, and as a means of transportation. In truth, however, the impetus to scientific study came from none of these things. It came from a passion for competitive camel racing on the part of Middle Eastern sheikhs, who have been known to pay more than $1 million for a superior racing camel, and who relish the prospect of a breeding program for camels similar to what has long existed for thoroughbred racehorses. Establishing such a program has turned out to be harder than anyone anticipated.

Lulu Skidmore works for His Highness Sheikh Mohammed bin Rashid al Maktoum, the crown prince of Dubai and the Defense Minister of the United Arab Emirates. The Maktoum family, which has ruled Dubai since the 1830s, has for decades been among the most powerful forces in the world of horse racing. Sheikh Mohammed himself owns more thoroughbred racehorses than anyone else in the world, and during a typical year in British competition the horses fielded by the Maktoum family's Godolphin Stables claim the greatest number of wins. Outside the Middle East the sheikh's deep interest in the ancient sport of camel racing is not well known. But the Maktoums are reputed to own a herd of 10,000 camels, and Sheikh Mohammed keeps a string of 2,000. His camels are eyed with envy by other sheikhs, and his stable maintains an imposing presence on the camel tracks of the Arab world.

The demands of camel racing have created, almost overnight, a thriving new field of biological endeavor — one that has proved irresistible even to researchers who began their careers with a different focus entirely. Two decades ago the field of camel biology was virtually nonexistent. For all the lore and mystique surrounding the camel, the byways of its physical functioning were far less well known than those of the cat, the rat, or the nematode. And yet by 1992 the First International Camel Conference could draw some two hundred specialists to Dubai. A second international conference is now being planned.

Lulu Skidmore is already looking beyond that point. "Obviously, the motivation for all this work was racing," she says. "But the sheikh is keen that Dubai should be on the map. From Dubai's point of view, it is important to be on the map in the world of science as well as in the world of camel racing. And the science could be doing a lot of good for other countries."

The desert sheikhdom of Dubai, one of the seven sheikhdoms that make up the United Arab Emirates, is essentially a city-state straddling a strategic inlet on the northeastern corner of the Arabian Peninsula, across the Persian Gulf from Iran. Oil was discovered in Dubai only three decades ago, and the oil will probably run out before another three decades have passed. The rulers of Dubai have set out to make the sheikhdom a mercantile entrepôt and financial hub, and they are succeeding. Skyscrapers in compelling shapes rise above Dubai Creek, where dhows are moored five deep along the waterfront. Construction is under way everywhere, and open trucks haul immigrant workers from site to site. In the older parts of the city, elegant wind towers of stone and daub rise above the houses, trapping the gulf breezes and directing them below. Desalination plants make possible golf courses and polo grounds and lush median strips. The souks are clean and freshly swept, and not far distant from shopping malls where the names Armani and Ralph Lauren and Moschino are prominent. There is no crime to speak of; the main local newspaper ran a story while I was there, earlier this year, about the trial of someone who had been charged with pickpocketing.

On the streets native Emiratis gowned in crisp white *dishdashas,* their heads covered with red-and-white-checked *gutras,* walk among a population that has become polychrome and polyglot. At the Hotel Inter-Continental the daily breakfast buffet features French toast, salted hammour fish, baked beans, channa masala, miso soup, fried rice, lamb chops, and Cheerios. Television channels are available in Japanese and Hindi. The one bookstore I found in Dubai had an aisle devoted to "Oprah's Book Club." Women in Dubai may drive cars, and Western women may lie on the beach in bikinis. But there is no lack of reminders about where you really are. On a table in my hotel room, a small sticker with an arrow labeled *QIBLA* showed the direction for prayer. Minarets rise above every neighborhood. A banner at an important intersection reminds the passing traffic, in Arabic and English, "Don't Forget Kuwait's Missing POWs."

The camel track lies on the outskirts of Dubai, at a place called Nad al Sheba, and during the racing season the green, red, white, and black national flag of the Emirates flutters atop high poles that line the road leading out from town. The races take place on a vast oval more than six miles around. In Saudi Arabia the races may go twice as far. Camels are not as fast as racehorses, but they have a lot more stamina — they start out at about 40 miles an hour, and can hold a pace of 20 miles an hour for more than half an hour. They have been known to run at 10 miles an hour for eighteen hours. The designation *dromedarius* has nothing to do with humps — it comes from the Greek word for "running." If the Greek were more candid, it would mean "running with a loopy gait while maintaining an expression of haughty insouciance."

During a race, in which fifty or more camels may compete, train-

ers and owners in jeeps speed around the inside of the track, talking with their jockeys by radio. The jockeys are helmeted boys,
each riding behind his camel's hump and wielding a riding crop
the length of a fishing rod. (Because weight makes a difference,
the age of jockeys has been an issue for years in some parts of the
camel-racing world; in Dubai the minimum weight was officially set
not long ago at about one hundred pounds, and authorities have
cracked down on offenders in several highly publicized recent
cases.) I watched scores of jockeys at play one afternoon as a dusky
blood-orange sun hovered above the sands and turned the boys
into lines of flickering shadow that seemed to stretch to the horizon. Behind them, camels by the hundreds were led by trainers
from the racetrack to the far stables, diminishing ultimately into little more than silhouetted flecks of movement. Looking at the sky, I
would not have been surprised to see the words "A David Lean Production."

Camel racing has always been a Bedouin pastime. But only during the past twenty years or so has it become a highly organized activity — the Sport of Sheikhs, as it might be called, with large in-
vestments in bloodlines and facilities. In the 1990s a dozen new
camel racetracks have been built in the Emirates alone. During the
month of March, camel racing is to Arabian television what basketball is to American. The grandstand at Nad al Sheba, under its tent-
like canopy, resembles a diminutive Denver airport. It is set among
date palms on a manicured lawn that covers the desert like a putting green. The crowd at a camel race is diverse: members of the
ruling family are thrown together with laborers from Pakistan and
Britons in blue blazers. No betting is permitted, because Islam
frowns on gambling. But a lottery dispenses door prizes among the
spectators, and the prizes carry the names Lexus and Range Rover.
At Nad al Sheba the morning races are primarily for camels owned
by sheikhs and the afternoon races are reserved for everyone else.
A Bedouin's dream is to win one of these afternoon races and sell
the victorious camel to a sheikh on the spot.

The intense interest in camel racing in the Emirates is not so
much about money as about cultural heritage. The traditional
Bedouin way of life is verging on extinction, but the animal at
its heart can still be exalted. It is difficult to overestimate the cen-
trality of the camel in the Arab imagination. The Koran praises

the horse and the sheep, but then comments pointedly, "The Almighty in making animals created nothing preferable to the camel." Speakers of Arabic have many words to express every refinement of the cameloid condition. *Dhaqun,* for instance, refers to "a she-camel that relaxes her chin so as to make her lower lip hang down while going along." Four-wheel drive has supplanted the camel's traditional function as a means of conveyance; nowadays you'll often see camels themselves being driven around in trucks. But people continue to hold camels in the highest esteem.

Inevitably, the fires of cultural nationalism have been fanned by the simoon of competition. The camel-racing sheikhs have ample resources, and they have sunk vast sums into the buying and training of camels. There are some 14,000 active racing camels in the Emirates. The sheikhs have built special treadmills and swimming pools to give the animals exercise. They have experimented with dietary supplements. They have mounted attacks on trypanosomosis and camel pox. They have hired physiologists, nutritionists, even psychologists. And, of course, they have begun to think about improving the racing stock, through scientific breeding programs and careful attention to pedigree.

Here the devotees of modern camel racing have faced a challenge. Think of the issue as a problem in mathematics. A thoroughbred racehorse starts competition at the age of two, and its career may be pretty well over by the age of four or five. It can begin life as a dam or a sire very early, and the track records of its progeny will be known within a few years — as will the track records of *their* progeny, and their progeny's progeny, and so on.

With camels, each step in this process is more time-consuming. A camel is not mature enough to start racing until the age of five or six, and its competitive abilities are not known until the age of six or seven. Scientific breeding can then start, but it will be another seven or eight years before the careers of the camel's first offspring can be evaluated.

Moreover, there won't be all that many offspring. In horse breeding a successful stallion can sire fifty to eighty foals a year, and may go to its grave with a thousand living descendants. But the preferred racing camels are female; the males are relatively ill tempered and difficult to deal with. Each pregnancy lasts thirteen months, and in a typical reproductive lifetime of twenty years, a camel may bear no more than about twelve calves. Here's an added element of torment for breeders: a camel's racing life potentially extends for many years beyond its reproductive maturity, and each pregnancy means the loss of more than a year of competition by what may be one of the fastest racing camels in Arabia.

There would be no way around these problems if the rules of thoroughbred horse racing applied to camels. Thoroughbreds must be conceived and brought to term the natural way — no shortcuts allowed, however enticing the options offered today by artificial insemination, embryo transfer, and other reproductive technologies. The traditionalism of thoroughbred breeding underlies the lucrative system of stud farms and the global cycling of prime stallions, who chase the hemispheric springtime from north to south, and in a good year may cover scores of mares apiece. But there is no Jockey Club in camel racing to lay down this kind of law, and there is plenty of money for science.

The desert in Dubai is as much a time as it is a place: it begins the moment resistance to it relents. The desert runs unbroken from Dubai City to the Hajar Mountains. Twenty-five miles southeast of Dubai, at Nakhlee, on the highway to Hatta, a gatepost sign you could easily miss marks an unpaved road to the right. Early one morning Lulu Skidmore pulled off the highway and onto this road. The sandy embankment on both sides was strewn with the yellow fruit known as desert apples. Geckos darted about. Down the road a little ways was a compound of low-slung ocher buildings. Camels stood in wire stockades everywhere: black camels from Saudi Arabia, tan camels from the Emirates, white camels from Sudan. In the stockades the females had been separated from the males. The males had been separated from one another.

This is the Camel Reproduction Centre, where Skidmore is the scientific director. The laboratory is a mainstay of Arabia's camel infrastructure — its Los Alamos, perhaps. Another mainstay is the Dubai Camel Hospital, near the Nad al Sheba racetrack, which serves as Skidmore's office. (One room in the hospital complex holds a raised slab with a large hole cut out of the middle. Skidmore responded to my raised eyebrow: "For the hump. It's an operating table.") Abu Dhabi, the sheikhdom adjacent to Dubai on the west, and the richest of the emirates, is home to three research stations: the Sheikh Hazza Camel Reproduction Research Centre, the Sheikh Khalifa Camel Embryo Transfer Research Centre, and the Sheikh Khalifa Scientific Centre for Racing Camels. The Saudis operate a Camel Research Centre at King Faisal University, in Al-Ahsa. Sultan Qaboos, of Oman, runs a small research facility in Muscat. All these places rely on imported staff and consultants, who have come from England and America, Australia and South Africa, and whose presence affirms the Koran's insight: Horses and sheep may be wonderful, but there is nothing preferable to the camel.

Skidmore was driving a Nissan Super Safari and talking on her ever-present cell phone. Its ring, a tinny version of "Whistle While You Work," serves as a kind of Lulu Positioning System. We drove past several men in long tunics and baggy pants — tribesmen from somewhere on the Afghan-Pakistan border. "The boys," Skidmore called them, or sometimes, more formally, "the camel guys." They looked vigilant and implacable. The camel guys were tethering a small herd of camels outside the entrance to the lab's examining bay, where the animals would be given ultrasound scans. We drove past a stocky young man who smiled and waved. This was Tipu Billah, who in effect manages the Camel Reproduction Centre. I have also seen him described, in some literature about the facility, as the person who "perfected the technique for collection of semen from camels for insemination purposes." We drove on a little farther, made a loop around the stockades, and came back to the compound. In a pen near the entrance, a vaguely llama-like creature kicked around a soccer ball, happily oblivious. "That's Rama," Skidmore said. I learned more about him later.

Inside, Skidmore showed off her lab — her incubators, her freezers, her progesterone-assay machine, her semen-loading machine — and introduced the lab technician, Ajaz Hussain. Then she donned beige coveralls, went out to the examining bay, and pulled a latex examination glove over her left arm. Ajaz Hussain handed her the ultrasound scanner.

For the next hour and a half, the camel guys orchestrated a decorous pavane. A camel would be led in and made to sit, its double-jointed hind legs collapsing tidily above a rope laid on the floor. The ends of the rope were pulled up and knotted atop the camel's back, immobilizing the animal. The camel guys put down a shovelful of feed where the camel could reach it. As the camel began to eat, Skidmore sat down on a mat near its other end, scooped out thirty or forty pellets of dry feces, and entered the rectum with the scanner until it reached a point just above the uterus, which offered sonic access to the ovaries. Skidmore's head was tilted up to the right, toward a video monitor on a low trolley. The grainy image on the monitor segued through wobbly mutations and then settled itself: Skidmore had found a follicle, and was counting eggs. She reported her findings aloud — "Another couple of days before she's at the right stage" — as Ajaz Hussain took notes. By then the camel guys had prepared another animal for examination. Skidmore withdrew the scanner, stood up, kicked the mat over to the next camel, sat down, and repeated the drill. She performed eight examinations, took a break, and performed eight more. This is how Skidmore spends every morning during the dromedary camel's reproductive season. The season runs from October to April.

Here is what Skidmore is up to. Using hormones, she wants to "superovulate your top-class female," as she puts it — that is, stimulate the very best racing camels to produce more than one egg at a time. (Her record is twenty-five.) She wants to fertilize those eggs, either by actually mating the female with a desired male or by means of artificial insemination. Eight days after fertilization, she wants to harvest all the embryos — she passes a balloon catheter through the cervix, inflates the balloon to seal the opening, fills the uterus with a saline solution, and then drains the solution, flushing the embryos out. Finally, she wants to transfer the microscopic embryos into the wombs of run-of-the-mill camels who can carry the embryos to term, allowing the biological mother to return to racing. To undertake these interventions, Skidmore needs to monitor every stage of every breeding camel's reproductive cycle with ultrasound scans and hormone assays. Getting to the point where clinical data even made any sense required tedious baseline studies, consuming Skidmore's first few years on the job. How do camels produce follicles? When do they show estrus? If they conceive, what are the hormonal changes? If they don't conceive, when do they come back in to estrus? There was nothing about any of this in the clinical literature.

During the break Skidmore sipped her coffee and gave some instructions to Tipu. Through the window I had noticed a female camel squatting expectantly in the sand, out by the stockades, and now one of the camel guys was leading a stallion toward her. The stallion was displaying his soft palate, a declaration of intent. From a shelf Tipu took what looked like a protective sheath for a telescope, a black rubber device about thirty inches long and five inches wide, open at one end and with a clear-plastic receptacle at the other. He tilted the opening toward me, displaying a precise configuration of foam rubber inside. "Very important to be just like the cervix," he explained. "Took many, many tries." He turned away and reached again for the shelf, retrieving a dented tube of K-Y lubricant.

Skidmore said to me, "I'll just be doing more of the same. Why don't you go with Tipu?"

Depending on your point of view, the dromedary camel is either the most improbable of animals or the only possible animal under the circumstances. Its split upper lip allows it to remove acacia leaves from among the spiky thorns — and yet its sharp teeth allow it to masticate thorny plants if it must. To shut out the blowing sand it has two sets of eyelashes, and nostrils that close tight. It can go for more than two weeks without drinking water, and when it does drink, it can tolerate salts and poisons that would kill a human being. The hump is not, of course, a water tank, as children sometimes think, though it does store a supply of fats that can be drawn down over time. What makes the camel resistant to heat and dehydration is its curious and still somewhat mystifying metabolism, which includes a thermal regulator that permits variations in body temperature of as much as eleven degrees in the course of a day. A camel filters out toxins so relentlessly and conserves water so stingily that it urinates only very small amounts. Arduous conditions may cause a camel's body mass to drop by 25 percent in a week — a decline that would prove fatal in other mammals. When water again becomes available, camels can drink an amount equal to a third of their weight in ten minutes. Oddly, this animal, so well adapted to places without water, is a more adept natural swimmer than the horse.

Sexually, the camel is decidedly out of the ordinary, as research by Skidmore and others has made clear. The editors of the

Proceedings of the First International Camel Conference, themselves well adapted to dry conditions, address the subject with nearly imperceptible whimsy. They observe, "Aspects of the reproductive mechanism appear to have been borrowed from many different genera such that it represents a strange 'hotch potch' of ideas and systems when viewed through the blinkered spectacles of the modern reproductive physiologist."

The female camel has an uneven uterus — it ovulates from both of the organ's horns but implants only in the left one. Like the female ferret and the female cat, the female camel is an "induced ovulator" — under natural conditions she will not actually release an egg until prompted to do so by the stimulus of mating. Male camels have their own peculiarities. A stallion can't bring himself to ejaculate in the vagina — he must be able to penetrate the very tight cervix. Hence the importance of Tipu's special foam-rubber lining. Perhaps owing to all the extra effort, male camels don't have the libido one might expect. Even at the height of the season they don't go around mating with everything in sight.

Lulu Skidmore came to camel research through her connection with the Equine Fertility Unit, a veterinary research facility in Newmarket, England, funded primarily by the Thoroughbred Breeders' Association. This is also, in a way, how Sheikh Mohammed came to camel research. The town of Newmarket, which lies about fifteen miles east of Cambridge, in Suffolk, has been the center of British racing since the days of Charles II. Its undulating heath supports scores of stud farms and pastures of the richest green, set off by tendrils of white fencing. An even brighter shade of green, in the distance, draws attention to the roof of Tattersalls', the bloodstock auction house. Around Newmarket everyone seems to have muddy boots and clear complexions, and a windswept flush on the cheeks. The racetrack at Newmarket Heath lies within the largest cultivated grassland in Europe; running right across the course are the remnants of the Devil's Ditch, a defensive earthenwork dating from early medieval times. This is what Lulu Skidmore's ancestors were building when Sheikh Mohammed's were inventing algebra.

As the Maktoum family moved heavily into horse racing, Sheikh Mohammed began buying stud farms in the Newmarket area. He now owns nine of them, including Dalham Hall, Hadrian, Rutland, Sommeries, and Dunchurch Lodge. With the Newmarket domains came proximity to the world of equine science, and specifically to the work of W. R. Allen. Allen, fifty-seven, is a professor of equine reproduction at Cambridge University and the director of the Equine Fertility Unit. The unit, which occupies 114 acres in a corner of the Duke of Sutherland's estate at Stetchworth, is devoted to studying reproductive problems in thoroughbred mares and stallions, an endeavor that frequently takes off in tangential directions. Its work is an outgrowth of the now-defunct Animal Research Station, in Cambridge, which pioneered embryo transfer in animals, and also pioneered semen freezing, embryo freezing, and embryo splitting. Known to everyone as Twink, Allen is a bluff, gregarious, straight-talking New Zealander. When I called to ask if I could come by, he consulted his calendar and said, "Monday at four for tea or Tuesday at six for gin." Strolling among the paddocks at dusk, after tea, he said matter-of-factly, "I want to be the first to clone a horse." Describing the challenge posed by the camel's cervix, he said, "Getting into it is pretty sporty."

When Sheikh Mohammed decided to pursue camel research, Twink Allen was an obvious person to consult. The sheikh knew the work of the Equine Fertility Unit — indeed, all the unit's hay barns and most of its fences had come secondhand from the sheikh when he decided to rebuild two of his nearby stud farms. Allen's son-in-law is Frankie Dettorri, the house jockey for the Maktoum family's Godolphin team. At the sheikh's invitation, Allen set out to create, in Dubai, something akin to an Equine Fertility Unit for camels. The effort proceeded slowly. For one thing, managing a facility in Dubai from a desk in Newmarket proved to be impractical. Moreover, the assumption that the camel was a horse with a hump turned out to be simplistic — as noted, all the basic science had to be done from scratch. Also, camels were not easy to handle: they could be stubborn and irascible, and proved intolerant of the poking and probing that cows and horses, standing upright, blithely accept. A single ultrasound scan might take hours.

Lulu Skidmore volunteered to take on the whole project. Skidmore had been what Allen calls the "embryo girl" at Newmarket — responsible for flushing and preserving embryos in non-thoroughbred breeding experiments — and had participated in the start-up work in Dubai. She had been an enthusiastic horsewoman all her life, had worked on a stud farm in South Africa, and had earned a degree in animal science from the University of London. She was independent and self-confident, with an adventurous streak. A senior Australian colleague, Professor Roger Short, advised Skidmore, as she later told me, "Look — forget horses. Everybody knows everything about horses. If I were you, I'd go into camels. Nobody knows a thing about camels. You could be the queen of the field." So she proposed herself for the job of "camel girl" — her term. Sheikh Mohammed's chief veterinary adviser, Ahmed Mustansir Billah, agreed to give it a try. The thought of working as a woman in an Arab culture did not bother Skidmore, and her hard work and obvious skill seem to have quelled any objections. Sheikh Mohammed himself underwrote her doctoral dissertation.

Her first decree as queen was to allow the camels to sit in her presence. No more upright palpation. It made all the difference. In our conversation Twink Allen kept drifting back to Skidmore's calming way with her subjects, describing it in a tone of awe. "She could pop in and scan the ovaries, which is what we needed to do, but that damn camel hardly knew she was there. I've seen them almost look around — 'Oh, it's you, Lulu. Good morning' — and carry on eating. When *we* were there, it sounded like a slaughterhouse with this banging and crashing, and camels moaning and groaning. Now it's like a morgue: shuffle, shuffle, shuffle, onto the ground, and away she goes and scans them. We designed some basic experiments to determine the reproductive physiology of the camel, which she completed and wrote up, and then she went on to do embryo transfer. Again, those small hands and the gentle female touch were helpful getting into the camel's vagina. Camels don't like it, whereas cows and horses don't give a damn."

Sheikh Mohammed endowed the Camel Reproduction Centre with a hundred racing camels and five stallions. Breeding experiments have brought the number to upwards of two hundred camels. Promising bloodlines can now be developed, though the crucible of the racetrack remains several seasons away. A few minutes before it was placed in a surrogate, I examined under a microscope an embryo flushed from a valuable racing camel identified only as No. 1805. The embryo looked unremarkable — a translucent circle with a bulbous rim. But in six years or so it may carry the sheikh's colors at Nad al Sheba.

"That was fast," Skidmore said when Tipu and I returned from outside. Tipu had taken up position at a strategic moment, engineering the interception. Now he held the black rubber tube by its neck, like a snake; the clear-plastic receptacle, at the other end, looked milky white. The storage of frozen sperm and frozen embryos has become routine for many animals, including horses and human beings, but camel sperm and camel embryos have so far proved largely uncooperative. Two visiting researchers at the lab, a woman from Cambridge University and a woman from Omaha's Henry Doorly Zoo, had some new techniques they wanted to try. They took Tipu's offering and went to work.

The camel research stations in Abu Dhabi and elsewhere have had their eye mainly on the racetrack finish line; one has begun to show interest in breeding endangered species. The research agenda of the Camel Reproduction Centre is considerably broader. Ahmed Mustansir Billah, who oversees the Maktoum faunal enterprises, and who is Skidmore's administrative superior, is insistent about that. I visited Billah one morning at his office on the grounds of the sprawling Zabeel Feed Mill, which prepares specially enhanced and highly secret comestibles for Sheikh Mohammed's quails and falcons, his ostriches and flamingos, his gazelles and foxes, and above all his horses and camels. The Zabeel Feed Mill is adjacent to Sheikh Mohammed's Zabeel Stables, and the Zabeel insignia — a white star on a maroon background — is prevalent. Bags of feed stamped with the Zabeel colors are moved by forklift. Employees in Zabeel jumpsuits scurry about. If this were a James Bond movie, a silo would open up and a white star would appear on the nose cone of a maroon missile.

Billah, who is Tipu's brother-in-law, proudly indicated a pile of books on top of a filing cabinet — copies of the camel conference's published *Proceedings*. In the lilting cadence of his native Pakistan, Billah's conversation ranged over many subjects relating to camels, including their personal qualities ("Such an affectionate animal — given a choice, I would always choose a camel over a horse") and their gastronomical potential ("If you go for a tender one, it is very delicious, with a slightly sharp taste"). But always he emphasized the importance of the basic science being done in Dubai: "In terms of scientific achievement we are far and away ahead of anyone else. Far and away. Our work is internationally recognized." Camel research, he wanted me to understand, isn't just for racing anymore.

The vaguely llama-like creature at the Camel Reproduction Centre, who spends his time kicking a soccer ball around, is a case in point. The story of Rama, as Skidmore had called him, began some 30 million years ago, when the two main camel genera started to separate. The family Camelidae had originated in North America. One branch eventually moved south, to the Andes, evolving into the genus *Lama,* which includes the guanaco (domesticated by the Incas into the modern llama) and the vicuña (domesticated into the alpaca). These New World camels became adapted to life at very high elevations and very low temperatures. The other branch of the family, which developed into the genus *Camelus,* moved north to Alaska and crossed into Siberia, eventually becoming the two-humped Bactrian camel of Mongolia and the one-humped dromedary camel that spread from northern India to Arabia to northern Africa. (It is a relative newcomer to Africa: the camel was unknown to the ancient Egyptians.) These Old World camels became adapted to living at sea level in hot, dry conditions.

Could the two branches be rejoined? They have the same number of chromosomes and the same reproductive oddities. But the Old World camels are by now six times as large as the New World ones. Inhabiting opposite sides of the equator, the two species also evolved so as to become reproductively active at different times of the year. Unlike mules, which are produced by unions of horses and donkeys, hybrids of camel and llama would never occur naturally. However, artificial insemination and hormone therapy open a world of possibility.

After many setbacks, Skidmore was able to impregnate a female guanaco with camel sperm, producing a *Camelus-Lama* hybrid, or "cama" — the first viable hybrid of Old and New World camels. Rama the cama, born on the day of the full moon during Ramadan, is so far the only creature of his kind. He has the woolly coat and the nose and nostrils of his mother, and the short ears and long tail of his father. His feet are a blend of the camel's two-toed conjoined footpad and the guanaco's cloven hoof. He is large but lacks a hump. Whether he is fertile will not be known for several years.

"It must be lonely, being him," I said to one of the frozen-sperm researchers, the young woman from Cambridge, as we watched Rama kick his soccer ball. She said, "Being a new species, you mean? He doesn't know it, though, does he? He thinks he's just like the rest of us." Behind us, in the examining bay, a camel's insistent moan acquired a tone of mild urgency as Skidmore pumped up a catheter balloon to plug the cervix. The camel guys grunted and leaned into the animal to keep it still. A tinny bar of "Whistle While You Work" somehow made itself heard.

The line on Sheikh Mohammed is that camel racing may have enticed him into the research field, but a broader view of camel science is why he stays. If the objective had been merely to learn how to pop out scores of healthy embryos from the very best camels, then the objective could be considered achieved, in Dubai and elsewhere. But at the Camel Reproduction Centre other investigations go on — into placentation, luteolysis, and cloning, for example. The ability to freeze camel semen and camel embryos probably won't have a big payoff at Nad al Sheba, but it could be crucial in revitalizing and improving genetic stocks elsewhere in the world, and in selecting for traits other than speed. Every year Skidmore brings in veterinarians from Kenya and Tunisia, from Mauritania and Ethiopia and elsewhere in the Third World, for a week-long training course — the Dubai equivalent of the Department of Agriculture's extension service. It's not entirely preposterous to imagine that we are in the process of creating a planet whose ecosystem will call rather urgently for what *Camelus dromedarius* has to offer.

Lulu Skidmore raised that possibility one afternoon as we sat in her office at the Dubai Camel Hospital. A room nearby holds a video library of all the camel races held at Nad al Sheba during the past decade, and I had watched enough for a lifetime. Through the half-open door of Skidmore's office, I had a view of the bleached camel skeleton that stands guard over the hospital lobby.

"We could do a lot of good for other countries where they really do need the camels for meat," Skidmore said. "Where they really do need them for milk. Where they desperately need them for transport. Worldwide, camels are becoming a much more important animal, as we kill off our environment by building everything up and draining the water out and pulling up trees. Before long a lot of the world is going to be desert — the desert is enlarging all the time. Camels will be one animal that can survive all this. We'll be farming camels instead of cattle and sheep. At the end of the day they're going to be a lifesaver."

Skidmore laughed. "Global warming," she said, "could be very good for me and my camels."

RICHARD PRESTON

The Demon in the Freezer

FROM *The New Yorker*

THE SMALLPOX VIRUS first became entangled with the human
species somewhere between 3,000 and 12,000 years ago — possibly in Egypt at the time of the pharaohs. Somewhere on earth at
roughly that time, the virus jumped out of an unknown animal into
its first human victim, and began to spread. Viruses are parasites
that multiply inside the cells of their hosts, and they are the smallest life forms. Smallpox developed a deep affinity for human beings. It is thought to have killed more people than any other infectious disease, including the Black Death of the Middle Ages. It
was declared eradicated from the human species in 1979, after a
twelve-year effort by a team of doctors and health workers from
the World Health Organization. Smallpox now exists only in laboratories.

Smallpox is explosively contagious, and it travels through the air.
Virus particles in the mouth become airborne when the host talks.
If you inhale a single particle of smallpox, you can come down with
the disease. After you've been infected, there is a typical incubation period of ten days. During that time, you feel normal. Then
the illness hits with a spike of fever, a backache, and vomiting, and
a bit later tiny red spots appear all over the body. The spots turn
into blisters, called pustules, and the pustules enlarge, filling with
pressurized opalescent pus. The eruption of pustules is sometimes
called the splitting of the dermis. The skin doesn't break but splits
horizontally, tearing away from its underlayers. The pustules become hard, bloated sacs the size of peas, encasing the body with
pus, and the skin resembles a cobblestone street.

The pain of the splitting is extraordinary. People lose the ability to speak, and their eyes can squeeze shut with pustules, but they remain alert. Death comes with a breathing arrest or a heart attack or shock or an immune-system storm, though exactly how smallpox kills a person is not known. There are many mysteries about the smallpox virus. Since the seventeenth century, doctors have understood that if the pustules merge into sheets across the body, the victim will usually die: the virus has split the whole skin. If the victim survives, the pustules turn into scabs and fall off, leaving scars. This is known as ordinary smallpox.

Some people develop extreme smallpox, which is loosely called black pox. Doctors separate black pox into two forms — flat smallpox and hemorrhagic smallpox. In a case of flat smallpox, the skin remains smooth and doesn't pustulate, but it darkens until it looks charred, and it can slip off the body in sheets. In hemorrhagic smallpox, black, unclotted blood oozes or runs from the mouth and other body orifices. Black pox is close to 100 percent fatal. If any sign of it appears in the body, the victim will almost certainly die. In the bloody cases, the virus destroys the linings of the throat, the stomach, the intestines, the rectum, and the vagina, and these membranes disintegrate. Fatal smallpox can destroy the body's entire skin — both the exterior skin and the interior skin that lines the passages of the body.

The smallpox virus's scientific name is *variola*. It means "spotted" in Latin, and it was given to the disease by a medieval bishop. The virus, as a life form, comes in two subspecies: *Variola minor* and *Variola major*. Minor is a weak mutant, and was first described in 1863 by doctors in Jamaica. People usually survive it. Classic major kills one out of three people if they haven't been vaccinated or if they've lost their immunity. The death rate with major can go higher — how much higher no one knows. *Variola major* killed half of its victims in an outbreak in Canada in 1924, and presumably many of them developed black pox. Smallpox is less contagious than measles but more contagious than mumps. It tends to go around until it has infected nearly everyone.

Most people today have no immunity to smallpox. The vaccine begins to wear off in many people after ten years. Mass vaccination for smallpox came to a worldwide halt around twenty-five years ago. There is now very little smallpox vaccine on hand in the

United States or anywhere else in the world. The World Health Organization (WHO) once had 10 million doses of the vaccine in storage in Geneva, Switzerland, but in 1990 an advisory committee recommended that most of it be destroyed, feeling that smallpox was no longer a threat. Nine and a half million doses are assumed to have been cooked in an oven, leaving the WHO with a total supply of 500,000 doses — one dose of smallpox vaccine for every 12,000 people on earth. A recent survey by the WHO revealed that there is only one factory in the world that has recently made even a small quantity of the vaccine, and there may be no factory capable of making sizable amounts. The vaccine was discovered in the age of Thomas Jefferson, and making a lot of it would seem simple, but so far the United States government has been unable to get any made at all. Variola virus is now classified as a Biosafety Level 4 hot agent — the most dangerous kind of virus — because it is lethal, airborne, and highly contagious, and is now exotic to the human species, and there is not enough vaccine to stop an outbreak. Experts feel that the appearance of a single case of smallpox anywhere on earth would be a global medical emergency.

At the present time, smallpox lives officially in only two repositories on the planet. One repository is in the United States, in a freezer at the headquarters of the federal Centers for Disease Control and Prevention, in Atlanta — the CDC. The other official smallpox repository is in a freezer at a Russian virology institute called Vector, also known as the State Research Institute of Virology and Biotechnology, which is situated outside the city of Novosibirsk, in Siberia. Vector is a huge, financially troubled former virus-weapons-development facility — a kind of decayed Los Alamos of viruses — which is trying to convert to peaceful enterprises.

There is a growing suspicion among experts that the smallpox virus may also live unofficially in clandestine biowarfare laboratories in a number of countries around the world, including labs on military bases in Russia that are closed to outside observers. The Central Intelligence Agency has become deeply alarmed about smallpox. Since 1995, a number of leading American biologists and public-health doctors have been given classified national-security briefings on smallpox. They have been shown classified evidence that as recently as 1992 Russia had the apparent capability of launching strategic-weapons-grade smallpox in special biological warheads on giant SS-18 intercontinental missiles that were targeted on the major cities of the United States. In the summer of last year, North Korea fired a ballistic missile over Japan in a test, and the missile fell into the sea. Some knowledgeable observers thought that the missile could have been designed to carry a biological warhead. If it had carried smallpox and landed in Japan, it could have devastated Japan's population: Japan has almost no smallpox vaccine on hand, and its government seems to have no ability to deal with a biological attack. The United States government keeps a list of nations and groups that it suspects either have clandestine stocks of smallpox or seem to be trying to buy or steal the virus. The list is classified, but it is said to include Russia, China, India, Pakistan, Israel, North Korea, Iraq, Iran, Cuba, and Serbia. The list may also include the terrorist organization of Osama bin Laden and, possibly, the Aum Shinrikyo sect of Japan — a quasi-religious group that has Ph.D. biologists as members and a belief that an apocalyptic war will bring them worldwide power. Aum members released nerve gas in the Tokyo subway in 1995, and, as the year 2000 approaches, the group is still active in Japan and in Russia. In any case, the idea that smallpox lives in only two freezers was never anything more than a comfortable fiction. No one knows exactly who has smallpox today, or where they keep the virus, or what they intend to do with it.

The man who is most widely credited with the eradication of smallpox from the human species is a doctor named Donald Ainslie Henderson. Everyone calls him D. A. Henderson. He was the director of the World Health Organization's Smallpox Eradication Unit from its inception, in 1966, to 1977, just before the last case occurred. "I'm one of many in the eradication," Henderson said to me once. "There's Frank Fenner, there's Isao Arita, Bill Foege, Nicole Grasset, Zdenek Jezek, Jock Copland, John Wickett — I could come up with fifty names. Let alone the tens of thousands who worked in the infected countries." Nevertheless, Henderson ran it. Smallpox killed at least 300 million people in the twentieth century. During that time, humanity was largely immune to smallpox, which is not the case today. When D. A. Henderson arrived on the scene in 1966, 2 million people a year were dying of smallpox.

In the years since the eradication effort began, Henderson and his team have effectively saved more than 50 million lives. This could be the most impressive achievement in the history of medicine. Henderson and his colleagues, however, have never received the Nobel Prize for their work.

D. A. Henderson is now a professor at the Johns Hopkins School of Public Health. He is the founder and the director of the Johns Hopkins Center for Civilian Biodefense Studies, a think tank that considers what might be done to protect the American population during a biological event. The term "biological event" hardly existed two years ago, but it is now used by emergency planners and by the FBI to mean a terrorist attack with a bioweapon — an unnatural event, caused by human intent.

Henderson lives with his wife in a large brick house in Baltimore. I arrived there on a cold, drizzly Saturday, and he ambled to the door. Henderson is an imposing man, six feet two inches tall. He is seventy years old. He has broad shoulders and a broad, seamed, angular face, pointed ears that stick out at angles from his head, a brush of gray hair, metal-framed eyeglasses, sharp blue eyes, and an easygoing voice that can flash with calculation. He was wearing a red checked shirt, with suspenders that held up Saturday slacks.

"In the last ten days, we've had fourteen different anthrax scares," he remarked in an offhand way as we stood in the front hallway of his house and he loomed over me. He has a top-secret-level national-security clearance, and he hears about little bio-terror events that don't get noticed by the media. He went on to say, "Everybody and his brother is threatening to use anthrax. This week, it happened in Atlanta, in Washington, D.C., in Michigan, and in California. It's largely hoaxes. Of course, a real bioterror event is going to happen one of these days."

We settled into easy chairs in the family room. The walls and shelves of the room were crowded with African and Asian sculptures and wooden Ethiopian crosses, which he had picked up in his global hunt for smallpox. A Japanese garden was visible through sliding-glass doors.

We ate ham and roast-beef sandwiches and drank Molson Ice beers. Henderson bit into a sandwich and chewed thoughtfully. Then he said, "Often, you get a worried look on your face, with the first signs of rash. We speak of the 'worried face' of smallpox. That face is a diagnostic sign. The rash comes up all at once. It's more dense on the face and the extremities. That's how you can tell smallpox from chicken pox. With chicken pox, the rash crops up over a period of days, and it's more dense on the chest and trunk of the body. Smallpox pustules have a dimple, a dent in the center. Doctors say that the pustules have a 'shotty' feel, like shotgun pellets. You can roll them between your fingertips under the skin."

"How many doctors could recognize smallpox today?" I asked.

"Virtually none. Smallpox takes forms that even I can't diagnose. And I wrote the textbook." He is a coauthor of *Smallpox and Its Eradication,* a large book in red covers, which experts call the Big Red Book of Smallpox. It was supposed to be the final word on smallpox — the tombstone of the virus.

On February 15, 1972, a thirty-eight-year-old Muslim clergyman returned to his home town of Damnjane, in Kosovo, Yugoslavia, after he'd been on a pilgrimage to Mecca, stopping at holy sites in Iraq. I will call him the Pilgrim. A photograph of the Pilgrim shows a man who looks well educated, has an intelligent face, and is wearing a clipped mustache and a beret. He had traveled by bus for his entire journey. The morning after his return home, he woke up feeling achy. At first he thought he was tired from the long bus ride, but then he realized that he had caught a bug. He shivered for a day or two and developed a red rash brought on by his fever, but quickly recovered. He had been vaccinated for smallpox two months earlier. Indeed, the Yugoslav medical authorities had been vaccinating the population of Yugoslavia relentlessly for more than fifty years, and the country was considered to be thoroughly immunized. The last case of smallpox in Yugoslavia had occurred in 1930.

The Pilgrim's family members and friends came to visit him. They wanted to hear about his trip, and he enjoyed telling them about it. Meanwhile, variola particles were leaking out of raw spots in the back of his throat and mixing with his saliva. When he spoke, tiny droplets of saliva, too small to be seen, drifted around him in a droplet cloud. If the person is throwing off a lethal virus, the cloud becomes a hot zone that can extend ten feet in all directions. Although the raw spots in the Pilgrim's throat amounted to a tiny surface of virus emission, smaller than a postage stamp, in a biological sense it was as hot as the surface of the sun, and it put enough smallpox into the air to paralyze Yugoslavia.

Variola particles are built to survive in the air. They are rounded-off rectangles that have a knobby, patterned surface — a gnarly hand-grenade look. Some experts call the particles bricks. The whole brick is made of a hundred different proteins, assembled and interlocked in a three-dimensional puzzle, which nobody has ever figured out. Virus experts feel that the structure of a smallpox particle is almost breathtakingly beautiful and deeply mathematical — one of the unexplored wonders of the viral universe. The structure protects the virus's genetic material: a long strand of DNA coiled in the center of the brick.

Pox bricks are the largest viruses. If a smallpox brick were the size of a real brick, then a cold-virus particle would be a blueberry sitting on the brick. But smallpox particles are still extremely small; about 3 million smallpox bricks laid down in rows would pave the period at the end of this sentence. A smallpox victim emits several bricks in each invisible droplet of saliva that spews into the air when the person speaks or coughs. When an airborne smallpox particle lands on a mucous membrane in someone's throat or lung, it sticks. It enters a cell and begins to make copies of itself. For one to three weeks, the virus spreads from cell to cell, amplifying silently in the body. No one has discovered exactly where the virus hides during its incubation phase. Probably it gets into the lymph cells, confusing the immune system, and victims are said to experience terrible dreams.

On February 21, when the Pilgrim had been feeling achy for almost a week, a thirty-year-old man, a schoolteacher, who is known to experts as Ljatif M., arrived in Djakovica, a few miles from the Pilgrim's town, to enroll in the Higher Institute of Education. Doctors who later investigated the schoolteacher's case never found out how he had come in contact with the Pilgrim. One of them must have ended up in the other's town. Possibly they stood next to each other in a shop — something like that.

On March 3 Ljatif developed a fever. Two days later, he went to a local medical center, where doctors gave him penicillin for his fever. Antibiotics have no effect on a virus. Then his skin broke with dark spots, and he may have developed a worried face. He felt worse, and a few days later his brother took him by bus to a hospital in the town of Cacak, about a hundred miles away. The dark spots were by this time merging into blackened, mottled splashes, which the doctors in Cacak didn't recognize. Ljatif became sicker. Finally, he was transferred by ambulance to Belgrade, where he was admitted to the Dermatology and Venereal Diseases Department of the city's main hospital. By then, his skin may have turned almost black in patches. We don't have access to his clinical reports, so I am describing a generalized extreme smallpox of the kind Ljatif had.

Inside the cells of the host, smallpox bricks pile up as if they were coming off a production line. Some of the particles develop tails. The tails are pieces of the cell's protein, which the virus steals from the cell for its own use. The tailed smallpox particles look like comets or spermatozoa. They begin to twist and wriggle, and they corkscrew through the cell, propelled by their tails toward the cell's outer membrane. You can see them with a microscope, thrashing with the same furious drive as sperm. They bump up against the inside of the cell membrane, and their heads make lumps, and the cell horripilates. Then something wonderful happens. Finger tubes begin to extend from the cell. The tubes grow longer. The cell turns into a Koosh ball. Inside each finger tube is a smallpox comet. The fingers lengthen until they touch and join nearby cells, and the smallpox comets squirm through the finger tubes into the next cell. The comets are protected from attack by the immune system because they stay inside the finger tubes, where antibodies and killer white blood cells can't reach them. Then the Koosh ball explodes. Out pour heaps of bricks that don't have tails. These smallpox particles are wrapped in a special armor, like hand grenades. They float away, still protected by their armor, and they stick to other cells and go inside them, and those cells turn into Koosh balls. Each infected cell releases up to 100,000 virus particles, and they are added to the quadrillions of particles replicating in the universe of the ruined host.

Ljatif's skin had become blackened, mottled, and silky to the touch, and sheets of small blood blisters may have peppered his face. In a case of black pox, variola shocks the immune system so that it can't produce pus. Small blood vessels were leaking and breaking in his skin, and blood was seeping under the surface. His skin had developed large areas of continuous bruises.

On March 9 the Belgrade doctors showed Ljatif to students and staff as a case that demonstrated an unusual reaction to penicillin. (In fact, a very bad reaction to penicillin can look like this.) Ljatif's eyes may have turned dark red. In hemorrhagic smallpox, one or two large hemorrhages appear in each eye, in the white encircling the iris, making the eyes look as if they could sag or leak blood. The eyes never do leak, but the blood in the eyes darkens, until the whites can sometimes seem almost black.

During the day of March 10, Ljatif suffered catastrophic hemorrhages into the intestines. His intestines filled up with blood, and he expelled quarts of it, staining the sheets black, and he developed grave anemia from blood loss. For some unknown reason, black-pox patients remain conscious, in a kind of paralyzed shock, and they seem acutely aware of what is happening nearly up to the point of death — "a peculiar state of apprehension and mental alertness that were said to be unlike the manifestations of any other infectious disease," in the words of the Big Red Book of Smallpox. We can imagine that Ljatif was extremely frightened and witnessed his hemorrhages with a sense that his insides were coming apart. During the final phase of a smallpox intestinal bleed-out, the lining of the intestines or the rectum can slip off. The lining is expelled through the anus, coming out in pieces or in lengths of tube. This bloody tissue is known as a tubular cast. When a smallpox patient throws a tubular cast, death is imminent. All we know about Ljatif is that his bleeds were unstoppable, that he was rushed to the Surgical Clinic of the Belgrade hospital, and that he died in the evening. The duty physician listed the cause of death as a bad reaction to penicillin.

"These hemorrhagic smallpox cases put an incredible amount of virus into the air," D. A. Henderson said. Some of the doctors and nurses who treated Ljatif were doomed. Indeed, Ljatif had seeded smallpox across Yugoslavia. Investigators later found that while he was in the hospital in Cacak he infected eight other patients and a nurse. The nurse died. One of the patients was a schoolboy, and he was sent home, where he broke with smallpox and infected his mother, and she died. In the Belgrade hospital, Ljatif infected twenty-seven more people, including seven nurses and doctors. Those victims infected five more people. Ljatif directly infected a total of thirty-eight people. They caught the virus by breathing the air near him. Eight of them died.

Meanwhile, the Pilgrim's smallpox traveled in waves through Yugoslavia. A rising tide of smallpox typically comes in fourteen-day waves — a wave of cases, a lull down to zero, and then a much bigger wave, another lull down to zero, then a huge and terrifying wave. The waves reflect the incubation periods, or generations, of the virus. Each wave or generation is anywhere from ten to twenty times as large as the last, so the virus grows exponentially and explosively, gathering strength like some kind of biological tsunami. This is because each infected person infects an average of ten to twenty more people. By the end of March 1972, more than 150 cases had occurred.

The Pilgrim had long since recovered. He didn't even know that he had started the outbreak. By then, however, Yugoslav doctors knew that they were dealing with smallpox, and they sent an urgent cable to the World Health Organization, asking for help.

Luckily, Yugoslavia had an authoritarian communist government, under Josip Broz Tito, and he exercised full emergency powers. His government mobilized the army and imposed strong measures to stop people from traveling and spreading the virus. Villages were closed by the army, roadblocks were thrown up, public meetings were prohibited, and hotels and apartment buildings were made into quarantine wards to hold people who had had contact with smallpox cases. Ten thousand people were locked up in these buildings by the Yugoslav military. The daily life of the country came to a shocked halt. At the same time, all the countries surrounding Yugoslavia closed their borders with it to prevent any travelers from coming out. Yugoslavia was cut off from the world. There were twenty-five foci of smallpox in the country. The virus had leapfrogged from town to town, even though the population had been heavily vaccinated. The Yugoslav authorities, helped by the WHO, began a massive campaign to revaccinate every person in Yugoslavia against smallpox; the population was 21 million. "They gave eighteen million doses in ten days," D. A. Henderson said. A person's immunity begins to grow immediately after the vaccination; it takes full effect within a week.

At the beginning of April, Henderson flew to Belgrade, where he found government officials in a state of deep alarm. The officials expected to see thousands of blistered, dying, contagious people streaming into hospitals any day. Henderson sat down with the minister of health and examined the statistics. He plotted the cases on a time line, and now he could see the generations of smallpox — one, two, three waves, each far larger than the previous one. Henderson had seen such waves appear many times before as smallpox rippled and amplified through human populations. Reading the viral surf with a practiced eye, he could see the start of the fourth wave. It was not climbing as steeply as he had expected. This meant that the waves had peaked. The outbreak was declining. Because of the military roadblocks, people weren't traveling, and the government was vaccinating everyone as fast as possible. "The outbreak is near an end," he declared to the minister of health. "I don't think you'll have more than ten additional cases." There were about a dozen: Henderson was right — the fourth wave never really materialized. The outbreak had been started by one man with the shivers. It was ended by a military crackdown and vaccination for every citizen.

At the present time, the United States' national stockpile of smallpox vaccine is a collection of four cardboard boxes that sit on a single pallet behind a chain-link fence inside a walk-in freezer in a warehouse in Lancaster County, Pennsylvania, near the Susquehanna River, at a facility owned by Wyeth-Ayerst Laboratories. The vaccine is slowly deteriorating. The Food and Drug Administration (FDA) has put a hold on the smallpox vaccine, and right now no one can use it — not even emergency personnel or key government leaders.

The vaccine is owned by the federal government and is managed by Wyeth-Ayerst, which is the company that made it, twenty-five to thirty years ago. It is stored in glass vials. The vials contain freeze-dried nuggets of live vaccinia virus. Vaccinia is a mild virus. When you are infected with it by vaccination, it causes a pustule to appear, and afterward you are immune to smallpox for some years. People who have been vaccinated have a circular scar the size of a nickel on their upper arm, left by the vaccinia-virus pustule they had in childhood after vaccination. Some adults can remember how much the pustule hurt.

People from Wyeth periodically open the boxes and send some of the vials out for testing, to see how the vaccine is doing. The vials once held 15 million good doses, but now moisture has invaded some of them. The nuggets are normally dry and white in color, but when moisture invades they turn brown and look sticky, and the vaccine may be weakened. The vaccine was made by a traditional method: The manufacturer had a farm where calves were raised. The calves' bellies were scratched with vaccinia virus, and their bellies developed pustules. Then the calves were killed and hung up on hooks, the blood was drained out of them, and the pustules were scraped with a knife. The resulting pus was freeze-dried. The vaccine is dried calf pus. According to one virologist who examined it under a microscope, "It looks like nose snot. It's all hair and wads of crap." It was a good vaccine for its time, but the FDA would never clear it for general use today, except in a national emergency. Furthermore, some people have bad or fatal reactions to the vaccine. There is an antidote, but the supplies of it have turned strangely pink, and the FDA has put a hold on the use of these supplies, too.

D. A. Henderson believes that in practice doctors could obtain about 7 million doses of vaccine from the vials. Unaccountably, most of the vaccine has not recently been tested for potency, so it has not been absolutely proved to work. The experts believe that it would work, but there still isn't enough.

Henderson explained the problem this way: "If there's a bioterror event, and someone releases enough smallpox to create a hundred cases — let's say in the Baltimore area — it would be a national emergency. The demand for vaccine would be beyond all belief." In Yugoslavia in 1972, the outbreak was started by one man, and 18 million doses of vaccine were needed — one for almost every person in the country.

"That first wave after the bioterror event could be a hundred people with smallpox," Henderson said. "It takes two weeks after exposure before doctors can diagnose smallpox. Meanwhile, those hundred people will give smallpox to a thousand or two thousand people. That's the second wave. Some of those first hundred people will go to other cities — to Washington, to New York, all over. So the second wave will include cases in other American cities, and probably in foreign countries. By then, it'll be too late to treat them, and we'll lose the second wave. We'll be well into the third wave — ten to twenty thousand people with smallpox — before we can really start vaccinating people. By then, we'll begin to pick up so many cases in the Baltimore area that we won't be able to track cases, and we'll just have to vaccinate everybody around Baltimore. A lot of people in Baltimore work in Washington. And so you're going to have a whole lot of people in Washington with smallpox. You can see the deal. Immediately, you would have to vaccinate Washington." Henderson thinks that 100 million doses of vaccine would be needed in the United States alone to stop a surging outbreak triggered by a hundred initial cases of smallpox from a bioterror event. That much vaccine could be stored in the space occupied by a one-car garage.

Raindrops splattered on a wooden deck in Henderson's garden, and the room grew dark, until it was a pool of shadows full of African masks. Henderson's voice came out of the gloom. He didn't bother to get up and turn on the lights. He said, "The way air travel is now, about six weeks would be enough time to seed cases around the world. Dropping an atomic bomb could cause casualties in a specific area, but dropping smallpox could engulf the world."

Henderson passionately wants to get rid of the virus. "What we need to do is create a general moral climate where smallpox is considered too morally reprehensible to be used as a weapon. That would make the possession of smallpox in a laboratory, anywhere, effectively a crime against humanity. The likelihood that it would be used as a weapon is diminished by a global commitment to destroy it. How much it is diminished I don't know. But it adds a level of safety."

In the late 1700s, the English country doctor Edward Jenner noticed that dairymaids who had contracted cowpox from cows seemed to be protected from catching smallpox, and he thought he would do an experiment. Cowpox (it probably lives in rodents, and only occasionally infects cows) produced a mild disease. On May 14, 1796, Jenner scratched the arm of a boy named James Phipps, introducing into the boy's arm a droplet of cowpox pus that he'd taken from a blister on the hand of a dairy worker named Sarah Nelmes. A few months later, he scratched the boy's arm with deadly pus he had taken from a smallpox patient, and the boy didn't come down with smallpox. The boy had become immune. Jenner had discovered what he called vaccination, after the Latin word for cow. He saw the road to eradication clearly. In 1801 he wrote, "It now becomes too manifest to admit of controversy, that the annihilation of the Small Pox, the most dreadful scourge of the human species, must be the final result of this practice."

A Soviet epidemiologist, Viktor Zhdanov, deserves credit for kick-starting the modern effort. At the 1958 annual meeting of the World Health Assembly, in Minneapolis, he called for the global eradication of smallpox. He spoke passionately and logically, but the scientific community was skeptical. Many biologists held a common view that it was impossible to separate a wild microorganism from the ecological web it lived in. In 1965 President Lyndon Johnson endorsed the idea of smallpox eradication. It was a political move to help improve American-Soviet relations. D. A. Henderson was then the head of disease surveillance at the CDC. He was given an order to report to Geneva to head the WHO's new Smallpox Eradication Unit. He didn't want the job, but he was told that if he didn't take it he would have to resign from government service. He went to Geneva, where he formed a hand-picked team. "The World Health Assembly proposed a ten-year program, because Kennedy had said we could land a man on the moon in ten years," he recalled.

The team set a goal of vaccinating 80 percent of the population of countries that harbored smallpox. Henderson says that from the beginning they had another idea as well, and it proved to be the key. The idea was to track smallpox outbreaks and vaccinate people in a ring around any outbreak. This is known as surveillance and ring vaccination. In order to throw a ring around smallpox, they had to know where the demon was moving at all times, and they started showing villagers photographs of a baby with smallpox so that the villagers could recognize and report cases to the authorities.

Henderson's team needed a way of vaccinating people fast. They tried a machine called the Ped-o-Jet, which was operated by foot pedal. It could shoot jets of vaccine into the arms of thousands of people in a day, but it broke down. Then they tried a needle with two points. It was known as the bifurcated needle, and it looked like a tiny two-pronged fork. The points of the fork held a droplet of vaccine, and the needle was to be jabbed repeatedly into a person's arm. It could be used by a volunteer who had no medical training.

They discovered that the virus rose and fell in seasonal waves, like flu. This led to an idea to attack the virus with a ring assault when it was at its ebb. The virus was a wild organism that lived only in humans. It needed to find and invade a susceptible human every fourteen days or it would die. If each outbreak of the virus could be surrounded by a ring of immune people during the virus's low season, the virus would not be able to complete its fourteen-day life cycle. It would be cut off, unable to move to the next human host, and its chain of infection would be broken.

The ring had to be tight. If it developed a leak, smallpox would blow out. In January 1975 smallpox blew out in Bangladesh, after the eradicators thought they were on the verge of stopping it everywhere in Asia. *Variola major* swept through more than five hundred towns and villages. Henderson began shuttling between Geneva and Bangladesh, and in April of that year, when things were still not under control, he visited the Infectious Disease Hospital in Dacca, the nation's capital. He wanted to do rounds in a smallpox ward. "I went down the rows of beds," he told me. There were seventy or eighty people, and half of them were dying. "There is nothing you can do for any of these patients. They were afraid to move. There were a lot of flies crawling all over the place. My God, they talk about the odor of smallpox. It *is* an odd smell, not like anything else."

The skin gives off gases. "It's a sickly odor, like rotting flesh, but it's not decay, because the skin remains sealed and the pus isn't leaking out," Henderson said. "That smell is one of the mysteries of smallpox. No one knows what it is. I was with this British guy, Nick Ward, M.D. He had worked in Africa — he was a tough guy. At the end, he stood by a fence looking at the ground. Finally, he said to me, 'I don't know that I could go through another situation like that again.'"

Nicholas Ward, who now lives in France, remembered that moment. "I've spent a fair amount of my life working with tropical diseases, and I can truly say there is nothing so awful as a case of smallpox, particularly the type where a person becomes a bloody mess," he said. He knew the odor. "I would have a shrewd idea of a diagnosis after walking into a home. I could smell it."

Henderson and his team mounted ring vaccinations across Bangladesh, and they traced cases and contacts, trying to surround the life form. Finally, in the fall of 1975, they cornered variola on an island off the coast of Bangladesh. It was a marshy, poor place called Bhola Island, and there, on October 16, a three-year-old girl named Rahima Banu broke with the last case of naturally occurring *Variola major* anywhere on earth. She survived. Rahima Banu would be twenty-seven years old today; researchers have lost track of her. Doctors from the Smallpox Eradication Unit collected six of the girl's pustules after they had dried into scabs, peeling them off her skin gently, with tweezers. Two years later, on October 26, 1977, the last natural case of the mild type of smallpox, *Variola minor,* popped up in a cook in Somalia named Ali Maow Maalin. He survived, and the last ring tightened around variola, and its life cycle stopped.

The headquarters of the Centers for Disease Control, in Atlanta, is a jumble of old and new buildings, joined by elevated walkways, which give the place the feel of a maze. The buildings sit along Clifton Road, an artery that winds through green neighborhoods in the northeastern part of Atlanta. I arrived at the CDC on a perfect day in spring. Changeable clouds marched across a deep sky, and oak trees were shedding green flowers. Across the street from the entrance, a blue jay screamed in a pine tree, and the branches glittered in the sun, throwing off a scent of pitch.

Joseph J. Esposito, Ph.D., who is the chief of the CDC's Poxvirus Section, led me along an outdoor walkway toward his laboratory and office. Esposito is a stocky man of moderate height, in his mid-fifties, who runs to keep his weight down, and he has a dark beard and wears eyeglasses over brown eyes that are perceptive and serious. I asked him if we could get closer to smallpox. We passed along an aerial walkway covered with a chain-link fence, and we turned onto another walkway. We stopped and leaned on a railing. We were facing the CDC's Level 4 biocontainment building. It contains the Level 4 hot suites — labs where researchers work with lethal viruses while wearing pressurized spacesuits. The building has a line of windows tinted blue-green, like fish tanks.

"The variola is in there somewhere," Joe Esposito said, offering me a grave smile and nodding at the Level 4 building. "There is a kind of electricity in the air when we're working with smallpox. Everybody around here always seems to know — 'Joe's got the smallpox out of the freezer.'"

The smallpox freezer may be encircled by alarms and motion detectors. It may or may not be wrapped in chains. It may be a stainless-steel cylinder. Or it may be a white box intended to look like any other freezer. Officials at the CDC won't comment.

Inside the freezer, the entire collection of smallpox occupies a volume slightly larger than that of a basketball. It consists of approximately four hundred little plastic vials the size of pencil stubs, the residue of D. A. Henderson's war with variola. They're an inch long and they have plastic screw caps. They sit in seven little white cardboard boxes in a rack inside the freezer, which keeps the virus not strictly alive, not exactly dead, but potent. Most of the vials contain milky ice or bloody ice. The virus has been cultured in flasks of live cells (milky ice) or in live chicken eggs (live eggs have a blood system). Around twenty-five of the vials contain human scabs — dried smallpox pustules. The scabs look like pencil erasers.

The six scabs that were collected from the girl in Bangladesh, named Rahima Banu, used to sit in a vial, but recently Esposito's group used the last of her scabs for research. The strain that came out of her scabs is known as Bangladesh 1975 — or, informally, as the Rahima. Now that the scabs are gone, the Rahima exists in vials of milky ice.

Esposito sat hunched in his chair in front of his computer. His office is a windowless room with cinder-block walls. A troll with shocking pink hair stood on top of the computer, staring at him wide-eyed. "I like to think like a virus," Esposito said. "If you can think like a virus, then you can begin to understand why a virus does what it does. A smallpox particle gets into a person's body and, in a way, it's thinking, 'I'm this one particle sitting here surrounded by an angry immune system. I have to multiply fast. Then I have to get out of this host fast.' It escapes into the air before the pustules develop." By the time the host feels sick, the virus has already moved on to its next host. The previous host has become a cast-off husk (and is now becoming saturated with virus), but whether the person lives or dies no longer matters to the virus. However, the dried scabs, when they fall off, contain live virus. The scabs are the virus's seeds. They preserve it for a long time, just in case it hasn't managed to reach a host in the air. The scabs give the virus a second chance.

Poxviruses move easily through the animal kingdom. Along with herds of animals or swarms of insects come poxviruses circulating among them like pickpockets at a fair. Esposito once classified what he and other virologists have glimpsed of the poxviruses in nature. He noted monkeypox, swinepox, buffalopox, skunkpox, raccoonpox, gerbilpox, a few deerpoxes, a sealpox, turkeypox, canarypox, pigeonpox, starlingpox, peacockpox, dolphinpox, Nile crocodilepox, penguinpox, two kangaroopoxes, and a quokkapox. (The quokka is an Australian wallaby.) Any attempt to get to the bottom of the butterflypoxes, mothpoxes, and beetlepoxes would be something like enumerating the nine billion names of God.

A caterpillar that has caught an insectpox dissolves into a lique-faction of insect guts mixed with pure crystals of poxvirus. This is known as a virus melt. The melt pours out of the dead caterpil-lar, and other caterpillars come along and accidentally eat the crys-tals lying on a leaf, and *they* melt, and so it goes for millions of years in the happy life of an insectpox. "It is a good thing no per-son has been known to catch an insectpox," Esposito remarked. (You might avoid eating melted caterpillars.) The yellow-fever mos-quito, *Aedes aegypti,* suffers from a fatal mosquitopox. At least two midgepoxes torment midges. Grasshoppers are known to get at least six poxes. If a grasshopperpox breaks out in a swarm of Afri-can locusts, it can wipe them out with a plague.

Viruses have an ability to move from one type of host to another in what is known as a trans-species jump. The virus changes during the course of a jump, adapting to its new host. The trans-species jump is the virus's most important means of long-term survival. Species go extinct; viruses move on. There is something impressive in the trans-species jump of a virus, like an unfurling of wings or a flash of stripes when a predator makes a rush. Some fifty years ago, in central Africa, the AIDS virus apparently moved out of chimpan-zees into people. Chimpanzees are now endangered, while the AIDS virus is booming.

For most of human prehistory, people lived in small groups of hunter-gatherers. The poxviruses did not deign to notice *Homo sa-piens* as long as the species consisted of scattered groups; there was no percentage in it for a pox. With the growth of agriculture, the human population of the earth swelled and became more tightly packed. Villages became towns and cities, and people were crowded together in river valleys.

Epidemiologists have done some mathematics on the spread

of smallpox, and they find that the virus needs a population of about 200,000 people living within a fourteen-day travel time from one another or the virus can't keep its life cycle going, and it dies out. Those conditions didn't occur in history until the appearance of settled agricultural areas and cities. At that point — roughly 7,000 years ago — the human species became an accident with a poxvirus waiting to happen.

Smallpox could be described as the first urban virus. It is thought to have made a trans-species jump into humans in one of the early agricultural river valleys — perhaps in the Nile Valley, or in Mesopotamia, or in the Indus River Valley. In the Cairo Museum, the mummy of the Pharaoh Ramses V, who died as a young man in 1157 B.C., is speckled with yellow blisters from face to scrotum.

In 1991 Joe Esposito and the molecular biologist Craig Venter, who was at the National Institutes of Health, sequenced the entire genome of the Rahima strain of smallpox; that is, they mapped all its DNA. They found that the virus contains 186,000 base pairs of DNA (each base pair being a step on the ladder of the molecule) and that the DNA contains about 187 genes — making smallpox one of the most complicated viruses known. (The AIDS virus has only 10 genes.) A gene is a piece of DNA, which contains the recipe for making one protein. Esposito's team noticed that smallpox has a gene that is also found in the placenta of a mouse. Smallpox knows how to make a mouse protein. How did smallpox learn that? "The poxviruses are promiscuous at capturing genes from their hosts," Esposito said. "It tells you that smallpox was once inside a mouse or some other small rodent." D. A. Henderson speculates that the original host of smallpox may have been an African rodent that lived in a crescent of green forests along the southern Nile River. The forests disappeared, cut down by people, and possibly the rodent has gone extinct. This is only a guess. Smallpox moved on.

The principal American biodefense laboratory is the United States Army Medical Research Institute of Infectious Diseases, or USAMRIID, in Fort Detrick, Maryland — an army base that nestles against the eastern front of the Appalachian Mountains in the city of Frederick, an hour's drive northwest of Washington. There is no smallpox at USAMRIID, for only the two WHO repositories are allowed to have it. The principal scientific adviser at

USAMRIID is Peter Jahrling, a civilian in his fifties with gray-blond hair, PhotoGray glasses, and a craggy face. Jahrling was the primary scientist during the 1989 outbreak of Ebola virus in Reston, Virginia: he discovered and named the Ebola-Reston virus.

"I don't think there is any higher biological threat to this nation than smallpox," Jahrling said to me, in his office, a windowless retreat jammed with paper. His voice was croaking. "I was over in Geneva for a meeting on smallpox, and I came back with some flu strain," he said hoarsely. The flu strain had swept through the world's smallpox experts. "Shows how fast a virus can move. If we have some kind of bioterror emergency with smallpox, there will be no time to start stroking our beards. We'd better have vaccine pre-positioned on pallets and ready to go."

Jahrling opposes the destruction of the official stocks of smallpox. "If you really believe there's a bioterrorist threat out there, then you can't get rid of smallpox," he said. "If smallpox is outlawed, only outlaws will have smallpox." His group has been testing antiviral drugs that might work on smallpox, and he feels that in order to verify the effectiveness of a new drug, it would be necessary to test it on live smallpox virus.

One of Jahrling's researchers, John Huggins, led me into the central areas of USAMRIID. Huggins is a chunky man with round Fiorucci eyeglasses. He turned into a corridor leading to the Level 4 spacesuit hot suites, or hot zones. The walls were cinder block, and the light turned bile green. A smoky reek drifted in the corridors, coming from huge autoclaves — pressure cookers — where contaminated equipment and waste were being heated and sterilized after being brought out of a hot zone. We stopped at a door that had a window of thick glass, looking into hot suite AA5, the Ebola hot zone.

I pressed my nose against the glass. It was cool, and there was a faint rumble of blowers, keeping the zone at negative pressure so that no contaminated air would flow out through cracks. The suite was dark and drowned in shadows, illuminated only by light coming from lab equipment. I could see no one in there but white mice in racks of plastic boxes. They were scribble-scrabbling in pine shavings.

"These mice are all infected with Ebola," Huggins said. "They bleed when they die. Like humans."

The mice looked fine. I couldn't see any blood in the shavings.

"We're giving them an antiviral drug that saves their lives," Huggins explained. "They're kind of perky. It's called an SAH drug. It's not ready for human testing. It could work in humans, but we don't know."

In 1995 Huggins spent time in a spacesuit at the CDC Level 4 lab in Atlanta, testing drugs on live smallpox. He found that a drug called cidofovir can block smallpox replication. Cidofovir, which is normally used against a virus that infects AIDS patients, has drawbacks. It must be given to people by IV drip, and there is some concern that it might damage the kidneys. Huggins and Jahrling believe that within five years better smallpox drugs are likely to be discovered. They say they will need to test the drugs directly on the virus. They add that the drug must be tested on the live virus in order to receive FDA approval.

In March a committee of the highly respected Institute of Medicine, in Washington, D.C., concluded that one of the main reasons for retaining live smallpox virus would be to help develop drugs against it. D. A. Henderson, who was not a member of the Institute of Medicine committee and thoroughly disagrees with its conclusions, thought that Jahrling was being too optimistic. "To get a new antiviral drug against smallpox is going to cost three hundred million dollars," he said. "The money simply isn't there."

Jahrling stood his ground. "Ceremonial destruction of smallpox is the crown jewel in D. A. Henderson's career," he said. "He would like to throw the lever on smallpox himself. If I had spent my life tramping the planet to eradicate the virus, I would want to throw the lever, too. What he did was a great accomplishment, but he has become blinded by the last glittering crown jewel of total eradication."

Ken Alibek, who was once Kanatjan Alibekov, a leading Soviet bioweaponeer and the inventor of the world's most powerful anthrax, shocked the American intelligence community when he defected, in 1992, and revealed how far the Soviet Union had gone with bioweapons. In a new book of his, entitled *Biohazard,* Alibek says that there were twenty tons of liquid smallpox kept on hand at Soviet military bases; it was kept ready for loading on biowarheads on missiles targeted on American cities. I contacted certain government sources and asked them if there was any evidence to corroborate Alibek's claims.

One person who asked not to be named said, "I really have to be careful what I say. Yeah, Alibek's claims have been corroborated in multiple ways. There's not a lot of evidence. There's some."

Another person who asked not to be named said that the Soviet Union had put the biowarheads on ICBM missiles and test-launched them sometime before 1991 over the Pacific Ocean. The United States — probably using spy satellites that orbited near the tests — was able to monitor the missiles as they soared into space and then punched back through the atmosphere and landed in the sea. The warheads were spinning weirdly: they were unusually heavy, and they had a strange shape. The warhead was heavy because it had an active refrigeration system to keep its temperature near or below the boiling temperature of water during reentry. Nuclear warheads don't need to be actively cooled. Why would a warhead need to be cooled? Presumably, because it was designed to contain something alive. But what? The person said, "The warhead was built to carry a very small quantity of biological weapon. Anthrax wouldn't have worked too well, because you need to put a lot of anthrax in the air to kill people, and anthrax isn't contagious. With smallpox, you don't need much. If you use smallpox, you get around the most difficult technical problem of bioweapons — the problem of dissemination. With smallpox, you use people as disseminators."

In 1989 a Soviet biologist named Vladimir Pasechnik defected to Britain. British intelligence agents spent a year debriefing him in a safe house. By the end, the British agents felt they had confirmed that the USSR had biological missiles aimed at the United States. This information reached President George Bush and the British prime minister, Margaret Thatcher. Mrs. Thatcher then apparently telephoned the Soviet leader, Mikhail Gorbachev, and sternly confronted him. She was furious, and so was Bush. Gorbachev responded by allowing a small, secret team of American and British biological-weapons inspectors to tour Soviet biowarfare facilities. In January of 1991 the inspectors traveled across the USSR, getting whirlwind looks at some of the major clandestine bases of the Soviet biowarfare program, which was called Biopreparat. The inspectors were frightened by what they discovered. ("I would describe it as scary, and I feel a responsibility to tell the world medical community about what I saw, because doctors could face these diseases," an inspector, Frank Malinoski, M.D., Ph.D., said to me.) On

January 14 the team arrived at Vector, the main virology complex, in Siberia, and the next day, after being treated to vodka and piles of caviar, they were shown into a laboratory called Building 6, where one of the inspectors, David Kelly, took a technician aside and asked him what virus they had been working with. The technician said that they had been working with smallpox. Kelly repeated the question three times. Three times, he asked the technician, "You mean you were working with *Variola major*?" and he emphasized to the technician that his answer was very important. The technician responded emphatically that it was *Variola major.* Kelly says that his interpreter was the best Russian interpreter the British government has. "There was no ambiguity," Kelly says.

The inspectors were stunned. Vector was not supposed to have any smallpox at all, much less be working with it. All the Russian smallpox stocks were supposed to be kept in one freezer in Moscow, which was supposed to be under the control of the World Health Organization. For Vector to have smallpox would be a supreme violation of rules set down by the WHO.

Then they went upstairs into Building 6 and entered a long corridor. On one side was a line of glass windows looking in on a giant airtight steel chamber of a type known as a dynamic aerosol test chamber. The device is for testing bioweapons. Small explosives are detonated inside the chamber, throwing a biological agent into the air of the chamber. The chamber in Building 6 had an octopuslike structure of tubes coming out of it, where sensors could be attached or monkeys could be clamped with their faces exposed to the chamber's air. An airborne bioweapon would get into the sensors or into the animals' lungs. On the other side of the corridor was a room that Frank Malinoski said "looked like a NASA control room," and video cameras provided views inside the chamber so that Vector scientists could watch the release of a bioweapon.

Vector scientists later told the inspectors that the chamber was a Model UKZD-25 — a bioweapons explosion-test chamber. It was the largest and most sophisticated modern bioweapons test chamber that has ever been found by inspectors in any country. It was used for testing smallpox.

The inspectors asked to put on spacesuits and to go inside. (They had brought along Q-Tip-like swab kits: they would have liked to swab the inner walls of the chamber, in the hope of collecting a virus.) The Russians refused. "They said our vaccines might not protect us," Malinoski says. "It suggested that they had developed viruses that were resistant to American vaccines." The Russians ordered the inspectors to leave Building 6.

At a large gathering that evening, three inspectors — David Kelly, Frank Malinoski, and Christopher Davis — publicly confronted the head of Vector, a virologist named Lev Sandakhchiev, about Vector's smallpox. (His name is pronounced "Sun-dock-chev.") He backpedaled angrily. Davis, a medical doctor with a Ph.D. who was then with British intelligence, now recalls, "Lev is gnomelike, a short man with a wizened, weather-beaten, lined face, and black hair. He's very bright and capable, a tough individual, full of bonhomie, but he can be very nasty when he is upset." Sandakhchiev heatedly insisted that his technician had misspoken. He called on his deputy, Sergei Netesov, to support him. The two Vector leaders insisted that there had been no work with smallpox at Vector. They had been doing genetic engineering with smallpox genes, they said, but Vector didn't have any live smallpox, only the virus's DNA — and the more they spoke the murkier their statements seemed. David Kelly remembers, "They were both lying, and it was a very, very tense moment. It seemed like an eternity, but it only lasted about fifteen minutes. And then there were so many other aspects of Vector we had to explore."

"The brazenness of these people!" one inspector later fumed. "They had been testing smallpox in their explosion chamber the week before we arrived."

Lev Sandakhchiev is still the head of Vector. He declined to be interviewed for this account but has steadfastly maintained that no offensive bioweapons research occurs now at Vector. In January of this year, at the Geneva meeting of smallpox experts, Sandakhchiev delivered a paper (and may have caught their flu). In his paper he claimed that Vector did not have any smallpox until 1994, when, he said, Vector had obtained it legally from Moscow. D. A. Henderson was also at that meeting. "It was quite elaborate and quite unbelievable," Henderson said. "I rolled my eyes, and saw other people rolling their eyes at me. We're sitting there, he's presenting us with all this horseshit, and he *knows* it's horseshit. Sandakhchiev is lying flagrantly."

Four sources have suggested to me that Lev Sandakhchiev was

in charge of a Vector research group that in 1990 devised a more efficient way to grow weapons-grade smallpox in industrial-scale pharmaceutical tanks known as bioreactors. The Vector smallpox bioreactors had a capacity of 630 liters — virus tanks big enough for a microbrewery. Once the Vector scientists had worked out the details of variola manufacturing, the results were written up in master production protocols — recipe books — and these protocols ended up at the Russian ministry of defense, in Moscow. At the time, weapons-grade smallpox was being manufactured by two older methods at a top-secret virus-munitions production plant near the city of Sergiyev Posad, forty-five miles northeast of Moscow. At another virus-munitions plant, near Pokrov, about two hundred miles southeast of Moscow, military virus-production specialists converted the plant to the new Vector method of making smallpox in the large virus bioreactors, but apparently never started the reaction. When one considers that a single person infected with smallpox would be considered a global medical emergency, this is rather a lot of smallpox activity to have bubbling near Moscow. It means that live smallpox virus and the protocols for how to mass-produce it had spread to various places in Russia by the 1990s. Indeed, live smallpox could be bubbling in reactors now at Sergiyev Posad — no one in the United States government admits to having a clue, and no Russian journalists have seen the place. Peter Jahrling said, "I really think that Vector is out of the offensive BW [biowarfare] business. But Sergiyev Posad is the black hole. We have no contacts there, and the Russians won't allow us to visit the place."

These days, Lev Sandakhchiev has cordial relationships with Peter Jahrling and Joe Esposito. They are eager to draw their colleague into the circle of open international science. During their visits to Siberia, Sandakhchiev has come across to them as warm and human, and desperate for research money to support his institute.

Sometimes, candid remarks slip out from the Russians. Jahrling put it this way: "There were tons of smallpox virus made in the Soviet Union. We know that. The Russians have admitted that to us. I was in a room with one of the Vector leaders when he said to us, 'Listen, we didn't account for every ampule of the virus. We had large quantities of it on hand. There were plenty of opportunities for staff members to walk away with an ampule. Although we think we know where our formerly employed scientists are, we can't account for all of them — we don't know where all of them are.'" Today, smallpox and its protocols could be anywhere in the world. A master seed strain of smallpox could be carried in a person's pocket. The seed itself could be a freeze-dried lump of virus the size of a jimmy on an ice-cream cone.

While I was sitting with D. A. Henderson in his house, I mentioned what seemed to me the great and tragic paradox of his life's work. The eradication caused the human species to lose its immunity to smallpox, and that was what made it possible for the Soviets to turn smallpox into a weapon rivaling the hydrogen bomb.

Henderson responded with silence, and then he said, thoughtfully, "I feel very sad about this. The eradication never would have succeeded without the Russians. Viktor Zhdanov started it, and they did so much. They were extremely proud of what they had done. I felt the virus was in good hands with the Russians. I never would have suspected. They made twenty tons — twenty tons — of smallpox. For us to have come so far with the disease, and now to have to deal with this human creation, when there are so many other problems in the world . . ." He was quiet again. "It's a great letdown," he said.

For years, the scientific community generally thought that biological weapons weren't effective as weapons, especially because it was thought that they're difficult to disperse in the air. This view persists, and one reason is that biologists know little or nothing about aerosol-particle technology. The silicon-chip industry is full of machines that can spread particles in the air. To learn more, I called a leading epidemiologist and bioterrorism expert, Michael Osterholm, who has been poking around companies and labs where these devices are invented. "I have a device the size of a credit card sitting on my desk," he said. "It makes an invisible mist of particles in the one- to five-micron size range — that size hangs in the air for hours, and gets into the lungs. You can run it on a camcorder battery. If you load it with two tablespoons of infectious fluid, it could fill a whole airport terminal with particles." Osterholm speculated that the device could create thousands of smallpox cases in the first wave. He feels that D. A. Henderson's estimate of how fast smallpox could balloon nationally is conservative. "D. A. is looking at Yugoslavia, where the population in 1972 had a lot of protective immunity," he said. "Those immune people are like control rods in a nuclear reactor. The American population has little immunity, so it's a reactor with no control rods. We could have an uncontrolled smallpox chain reaction." This would be something that terrorism experts refer to as a "soft kill" of the United States of America.

The idea that a biological credit card could execute a soft kill of the United States has reached the White House. The chief terrorism expert on the National Security Council, Richard Clarke, has sent word through the federal government that getting national stockpiles of smallpox vaccine is a top priority.

The effort started four years ago. So far, the government has little to show except numerous meetings among agencies, with no hope of vaccine anytime soon. The Department of Defense has put all its vaccine efforts into something called the Joint Vaccine Acquisition Program, which is run by the Joint Program Office for Biological Defense. People inside the military don't want their names used when they talk about the Pentagon's efforts. "It's a fucking disaster," said one knowledgeable military officer who has had direct experience in the matter. Last year the Pentagon hired a systems contractor called Dynport, headquartered in Reston, Virginia, to develop and make a number of different vaccines for troops. The smallpox-vaccine contract calls for 300,000 doses, at a cost of $22.4 million, or $75 a dose, with delivery now scheduled for 2006. (The date has been pushed back at least once already.) This amount of vaccine could be made in about fifteen flasks the size of soda bottles. There are 2.3 million people in the armed forces, and they have several million more dependents. "Three hundred thousand doses is not enough vaccine to protect anyone — not even our troops. It totally ignores the fact that smallpox is contagious," one military man said. "These guys ought to be buying tank treads and belt buckles. They know nothing about vaccines."

The Department of Health and Human Services (HHS) has been given the responsibility by the White House for producing a stockpile of smallpox vaccine large enough to protect the American civilian population in case of a bioterror event; originally, the idea was for HHS to consider hiring the military's contractor, Dynport, to make 40 million extra doses, in addition to the

300,000 that Dynport was making for the Pentagon. (Any such initiative would require competitive bidding.)

At a series of meetings at HHS, a top Dynport executive said that 40 million doses could be quite expensive. One scientist asked if a group of knowledgeable people could be drawn together to come up with an estimate of costs. The Dynport man answered, "Yes, we can do a study that will list the questions that need to be asked. It will cost two hundred and forty thousand dollars and will take six weeks."

Somebody then asked how much it would cost to *answer* the questions. The Dynport official responded, "That will be a different study. That study will cost two million dollars and will take six months."

With that, one scientist at the meeting burst out, "This is horseshit! We're asking an encyclopedia salesman if we need an encyclopedia!"

The CEO of Dynport, Stephen Prior, said that the situation is more complicated: "The civilian population is very different from the military. There's an age spread from newborns to the elderly; there's more compromised immunity, with AIDS, chemotherapy, and organ transplants. And possibly thirty-five percent of people have never been vaccinated. So it's not just scaling up the manufacturing."

Another knowledgeable observer is the retired army general Philip K. Russell, M.D., who gave the order to send biohazard troops into Reston in 1989 to deal with a building full of monkeys infected with Ebola. Russell said to me, "Many of us are afraid that Dynport won't deliver the goods without wasting an inordinate amount of money."

However, HHS has quietly opened talks with other potential contractors, preparing to solicit bids to make a civilian stockpile of smallpox vaccine, though there has been no announcement. "The effort at HHS still isn't organized," D. A. Henderson said. General Russell said, "If smallpox really got going, people should be most concerned about a lack of effective leadership on the part of their government."

I wanted to get closer to smallpox virus. In Joe Esposito's lab, at the Centers for Disease Control, there was a test going on of a biosensor device for detecting smallpox. It was a machine in a black suit-case. It could detect a bioweapon using the process called the poly-merase chain reaction, or PCR — the same kind of molecular fingerprinting that police use to identify the DNA of a crime sus-pect. The suitcase thing was called a Cepheid Briefcase Smart Cycler, and it had been coinvented by M. Allen Northrup, a bio-medical engineer who founded a company to make and sell biosensors. He was there, along with a cluster of other scientists.

Esposito, the official guardian of one half of the world's official supply of smallpox, handed a box of tubes to a scientist in the room. Two of the tubes contained the whole DNA of smallpox virus but not live smallpox. The DNA drifted in a drop of water; it was the Rahima strain. Two other tubes contained anthrax. The sam-ples were snapped into slots in the machine.

Northrup turned his attention to a laptop computer that nestled in the machine. Northrup is a chunky man with a mustache and reddish brown hair. He tapped on the keys.

We waited around, chatting. Meanwhile, the Cepheid was work-ing silently. It showed colored lines on its screen. In fifteen min-utes, the anthrax lines started going straight up, and someone said, "The anthrax is screaming." Finally, one of the smallpox lines crept upward, slowly. "That's a positive for smallpox, not so bad," a scien-tist said. Emergency-response teams could carry a Cepheid suitcase to the scene of a bioterror event and begin testing people immedi-ately for anthrax or smallpox. The machine is priced at $60,000.

Afterward, Joe Esposito went around collecting the used tubes. The smallpox-sample holder — a plastic thing the size of a thumb-nail — had been left on a counter. I picked it up.

Esposito wasn't about to let anyone walk off with smallpox. "Leave me that tube," he said. "You are not allowed to have more than twenty percent of the DNA."

Before I handed it to him, I glanced at a little window in the tube. When I held it up to the light, the liquid looked like clear wa-ter. The water contained the whole molecules of life from variola, a parasite that had colonized us thousands of years ago. We had al-most freed ourselves of it, but we found we had developed a strong affinity for smallpox. Some of us had made it into a weapon, and now we couldn't get rid of it. I wondered if we ever would, for the story of our entanglement with smallpox is not yet ended.

OLIVER SACKS

Brilliant Light

FROM *The New Yorker*

Something has got into me these last weeks — I do not know why. I have pulled out my old books (and bought many new ones), have set the little tungsten bar on a pedestal and papered the kitchen with chemical charts. I read lists of cosmic abundances in the bath. On cold, dismal Saturday afternoons, there is nothing better than curling up with a fat volume of Thorpe's Dictionary of Applied Chemistry, *opening it anywhere, and reading at random. It was Uncle Tungsten's favorite book, and now it is one of mine. On depressive mornings, I like to work out atomic radii or ionization potentials with my Grape-Nuts — their charm has come back, and they will get me going for the day.*

I — Before the War

Many of my childhood memories are of metals: these seemed to exert a power on me from the start. They stood out, conspicuous against the heterogeneousness of the world, by their shining, gleaming quality, their silveriness, their smoothness and weight. They seemed cool to the touch, and they rang when they were struck.

I also loved the yellowness, the heaviness of gold. My mother would take the wedding ring from her finger and let me handle it for a while, as she told me of its inviolacy, how it never tarnished. "Feel how heavy it is," she would add. "It's even heavier than lead." I knew what lead was, for I had handled the heavy, soft piping the plumber had left behind one year. Gold was soft, too, my mother told me, so it was usually combined with another metal to make it harder.

It was the same with copper — people mixed it with tin to produce bronze. Bronze! The very word was like a trumpet to me, for battle was the brave clash of bronze upon bronze, bronze spears on bronze shields, the great shield of Achilles. Or you could alloy copper with zinc, my mother said, to produce brass. All of us — my mother, my brothers, and I — had our own brass menorahs for Hanukkah. (My father, though, had a silver one.)

I knew copper, the shiny rose color of the great copper cauldron in our kitchen — it was taken down only once a year, when the quinces and crab apples were ripe in the garden and my mother would stew them to make jelly.

I knew zinc — the dull, slightly bluish birdbath in the garden was made of zinc — and tin, from the heavy tinfoil in which sandwiches were wrapped for a picnic. My mother showed me that when tin or zinc was bent it uttered a special "cry." "It's due to deformation of the crystal structure," she said, forgetting that I was five and could not understand her — and yet her words made me want to know more.

There was an enormous cast-iron lawn roller out in the garden — it weighed five hundred pounds, my father said. We, as children, could hardly budge it, but he was immensely strong and could lift it off the ground. It was always slightly rusty, and this bothered me, for the rust flaked off, leaving little cavities and scabs, and I was afraid the whole roller might corrode and fall apart one day, reduced to a mass of red dust and flakes. I needed to think of metals as stable, like gold — able to stave off the losses and ravages of time.

I would sometimes beg my mother to bring out her engagement ring and show me the diamond in it. It flashed like nothing I had ever seen, almost as if it gave out more light than it took in. My mother showed me how easily it scratched glass, and then she told me to put it to my lips. It was strangely, startlingly cold — metals felt cool to the touch, but the diamond was icy. That was because it conducted heat so well, she said — better than any metal — so it drew the body heat away from one's lips when they touched it. This was a feeling I was never to forget. Another time, she showed me how if one touched a diamond to a cube of ice, it would draw heat from one's hand into the ice and cut straight through it as if it were butter.

My mother told me that diamond was a special form of carbon, like the coal we used in every room in winter. I was puzzled by this — how could black, flaky, opaque coal be the same as the hard, transparent gemstone in her ring?

I would gaze into the heart of the coal fire, watching it go from a dim red glow to orange, to yellow, and blow on it with the bellows until it glowed almost white-hot. If it got hot enough, I wondered, would it blaze blue, be blue-hot?

I loved light, especially the lighting of the Sabbath candles on Friday nights, when my mother would murmur a prayer as she lit them. I was not allowed to touch them once they were lit — they were sacred, I was told, their flames were holy, not to be fiddled with. The little cone of blue flame at the candle's center — why was it blue? Did the sun and stars burn in the same way? Why did they never go out?

My mother showed me other wonders — she had a necklace of polished yellow pieces of amber, and she showed me how, when she rubbed them, tiny pieces of paper would fly up and stick to them. Or she would put the electrified amber against my ear, and I would hear and feel a tiny snap, a spark.

My older brothers Marcus and David were fond of magnets and enjoyed demonstrating them to me, drawing a magnet beneath a piece of paper on which were strewn powdery iron filings. I never tired of the remarkable patterns which rayed out from the poles of the magnet. "Those are lines of force," my brother Marcus explained to me — but I was none the wiser.

Then there was the crystal radio I played with in bed, jiggling the wire on the crystal until I got a station loud and clear. And the luminous clocks — the house was full of them because my Uncle Abe had been a pioneer in the development of luminous paints. These, too, like my crystal radio, I would take under the bedclothes at night, into my private, secret vault, and they would light up my cavern of sheets with an eerie, greenish light.

All these things — the rubbed amber, the magnets, the crystal radio, the clock dials with their tireless coruscations — gave me a sense of invisible rays and forces, a sense that beneath the familiar world of colors and appearances lay a dark, hidden world of mysterious laws and phenomena.

*

I grew up in London, before the war. My father and mother were both physicians. My father had his surgery in the house, with all sorts of medicines, lotions, and elixirs in the dispensary — it looked like an old-fashioned chemist's shop in miniature — and a small lab with a spirit lamp, test tubes, and reagents for testing patients' urine, like the bright-blue Fehling's solution, which turned yellow when there was sugar in the urine. There were potions and cordials in cherry red and golden yellow, and colorful liniments like gentian violet and malachite green.

I badgered my parents constantly with questions. Where did color come from? Why did the platinum loop cause the gas to catch fire? What happened to the sugar when one stirred it into the tea? Where did it go? Why did water bubble when it boiled? (I liked to watch water set to boil on the stove, to see it quivering with heat before it burst into bubbles.)

Whenever we had "a fuse," my father would climb up to the porcelain fuse box high on the kitchen wall, identify the fused fuse, now reduced to a melted blob, and replace it with a new one of an odd, soft wire. I had not known that a metal could melt, nor did I know why it had melted. Could a fuse really be made from the same material as a lawn roller or a tin can?

What was electricity, and how did it flow? Was it a sort of fluid like heat, which could also be conducted? Why did it flow through the metal but not the porcelain? This, too, called for explanation.

My questions were endless, and touched on everything, though they tended to circle around, return to, my obsession, the metals. Why were they shiny? Why smooth? Why cool? Why hard? Why heavy? Why did they bend, not break? Why did they ring? My mother tried to explain, but eventually, when I exhausted her patience, she would say, "That's all I can tell you — you'll have to quiz Uncle Dave to learn more."

We had called him "Uncle Tungsten" for as long as I could remember, because he manufactured light bulbs with filaments of fine tungsten wire. (His firm was called Tungstalite.) I often visited him in his old factory in Farringdon and watched him at work, in a wing collar, with his shirtsleeves rolled up. The heavy dark tungsten powder would be pressed, hammered, sintered at red heat, then drawn into finer and finer wire for the filaments. Uncle's hands were seamed with the black powder, beyond the power of any washing to get out. After thirty years of his working with tungsten, I imagined, the heavy element was in his lungs and bones, in every vessel and viscus, every tissue of his body. I thought of this as a wonder, not a curse — his body invigorated and fortified by the mighty element, given a strength and enduringness almost more than human.

Whenever I visited the factory, he would take me around the machines, or have his foreman do so. (The foreman was a short, muscular man, a Popeye with enormous forearms, a palpable testament to the benefits of working with tungsten.) I never tired of the ingenious machines, always beautifully clean and sleek and oiled, or the furnace where the black powder was compacted from a powdery incoherence into dense, hard bars with a gray sheen.

During my visits to the factory, and sometimes at home, Uncle Dave would teach me about metals with little experiments. I knew that mercury, that strange liquid metal, was incredibly heavy and dense. Even lead floated on it, as my uncle showed me with a lead bullet and a bowl of quicksilver. But then he pulled out a small gray bar from his pocket, and, to my amazement, it sank immediately to the bottom. This, he said, was his metal: tungsten.

Uncle loved the density of the tungsten he made, and its refractoriness, its great chemical stability. He loved to handle it — the wire, the powder, but the massy little bars and ingots most of all. He caressed them, balanced them (tenderly, it seemed to me) in his hands. "Feel it, Oliver!" he would say, thrusting a bar at me. "Nothing in the world feels like sintered tungsten." He would tap the little bars and they would emit a deep clink. "The sound of tungsten," Uncle Dave would say. "Nothing like it." I did not know whether this was true, but I never questioned it.

II — Exile

In September 1939, war broke out. It was expected that London would be heavily bombed, and parents were pressured by the government to evacuate their children to safety in the countryside. My brother Michael, five years older than I, had been going to a day school near our house in West Hampstead, and when it was closed at the outbreak of the war, one of the assistant masters there decided to reconstitute the school in a little village I will call Grey-

stone. My parents (I was to realize many years later) were greatly
worried about the consequences of separating a little boy — I was
just six — from his family and sending him to a makeshift board-
ing school in the Midlands, but they felt they had no choice,
and took some comfort that at least Michael and I would be together.

This, perhaps, might have worked out well — evacuation did
work out reasonably well for thousands of others. But the school,
as reconstituted, was a travesty of the original. Food was rationed
and scarce, and our food parcels from home were looted by the
matron. Our basic diet was swedes and mangelwurzels — giant tur-
nips and huge, coarse beetroots grown basically for cattle. There
was a steam-pudding whose revolting, suffocating smell comes
back to me (as I write, almost sixty years later) and sets me retching
and gagging once again. The horribleness of the school was made
worse for most of us by the sense that we had been abandoned by
our families, left to rot in this awful place as an inexplicable punishment for something we had done.

The headmaster seemed to have become unhinged by his own
power. He had been decent enough, even well liked, as a teacher in
London, Michael said, but at Greystone, where he took over, he
had quickly become a monster. He was vicious and sadistic, and
beat many of us almost daily, with relish. "Willfulness" was severely
punished. I sometimes wondered if I was his "darling," the one selected for a maximum of punishment, but in fact many of us were
so beaten we could hardly sit down for days on end. Once, when he
had broken a cane on my eight-year-old bottom, he roared, "Damn
you, Sacks! Look what you have made me do!" and added the cost
of the cane to my bill. Bullying and cruelty, meanwhile, were rife
among the boys, and great ingenuity was exercised in finding out
the weak points of the smaller children and tormenting them beyond bearable limits. I felt trapped at Greystone, without hope,
without recourse, forever — and many of us, I suspect, were severely disturbed by being there.

And yet the old vicarage, with its spacious garden, where the
school was housed, the old church next door to it, the village itself,
and the countryside surrounding it were charming, even idyllic.
The villagers were kind to these obviously uprooted and unhappy
young boys from London. It was here in the village that I learned to
ride horses, with a strapping young woman; she sometimes hugged

me when I looked miserable. (My brother had read me parts of *Gulliver's Travels,* and I sometimes thought of her as Glumdalclitch, Gulliver's giant nurse.) There was an old lady to whom I went for piano lessons, and she would make tea for me. And there was the village shop, where I would go to buy gob-stoppers and, occasionally, a slice of Spam. There were even times in school that I enjoyed: making model planes of balsa wood, and a tree house with a friend, a red-haired boy of my own age.

During the four years I was at Greystone, my parents visited us at the school, but very rarely. There was one return visit to London, in December of 1940 — a frightening one, because the Blitz was still at its height. One night a thousand-pound bomb fell on the house next to ours. Fortunately, it failed to explode. All of us, the entire street, it seemed, crept away that night (my family to a cousin's house) — many of us in our pajamas — walking as softly as we could. (Might vibration set the thing off?) The streets were pitch dark, for the blackout was in force, and we all carried electric torches dimmed with red crepe paper. We had no idea if our houses would still be standing in the morning.

On another occasion, an incendiary bomb, a thermite bomb, fell behind our house and burned with a terrible, white-hot heat. My father had a stirrup pump, and my brothers carried pails of water to him, but water seemed useless against this infernal fire — indeed, made it burn even more furiously. There was a vicious hissing and sputtering when the water hit the metal, and meanwhile the bomb was melting its casing and throwing white-hot blobs and jets of molten metal in all directions. The lawn was as scarred and charred as a volcanic landscape the next morning, but littered, to my delight, with beautiful gleaming shrapnel. I had never seen melted iron or magnesium before.

There had been some religious feeling, of a childish sort, in the years before the war. When my mother lit the Sabbath candles, I would feel, almost physically, the Sabbath coming in, being welcomed, descending like a soft mantle over the earth. I imagined, too, that this occurred all over the universe — the Sabbath descending on far-off star systems and galaxies, enfolding them all in the peace of God.

But when I was suddenly abandoned by my parents (as I saw it), my trust in them, my love for them, was rudely shaken, and with this my belief in God, too. What evidence was there, I kept asking myself, for God's existence? I determined on an experiment to resolve the matter decisively: I planted two rows of radishes side by side in the vegetable garden and asked God to bless one or curse one, whichever he wished, so that I might see a clear difference between them. The two rows of radishes came up identical, and this was proof for me that no God existed. But I longed now even more for something to believe in.

As the beatings, the starvings, the tormentings continued, those of us who remained at school (many had been taken away by their parents, but Michael and I never complained) were driven to more and more extreme psychological measures — dehumanizing, derealizing, our chief tormentor. Sometimes, while being beaten, I would see him reduced to a gesticulating skeleton. (I had seen radiographs at home — bones in a tenuous envelope of flesh.) At other times, I would see him as not a being at all but a temporary vertical collection of atoms. I would say to myself, "He's only atoms" — and, more and more, I craved a world that was "only atoms." The violence exuded by the headmaster seemed at times to contaminate the whole of living nature, so that I saw violence as the very principle of life.

What could I do, in these circumstances, other than seek a private place, a refuge where I might be alone, absorb myself without interference from others, and find some sense of stability and warmth? My situation was perhaps similar to that which Freeman Dyson describes in his autobiographical essay "To Teach or Not to Teach":

> I belonged to a small minority of boys who were lacking in physical strength and athletic prowess . . . and squeezed between the twin oppressions of whip and sandpaper [a vicious headmaster and bullying boys]. . . . We found our refuge in a territory that was equally inaccessible to our Latin-obsessed headmaster and our football-obsessed schoolmates. We found our refuge in science. . . . We learned . . . that science is a territory of freedom and friendship in the midst of tyranny and hatred.

For me, the refuge I found at first was in numbers. My father was a whiz at mental arithmetic, and I, too, even at the age of six, was quick with figures — and, more, in love with them. I liked numbers because they were solid, invariant; they stood unmoved in a chaotic world. There was in numbers and their relation something absolute, certain, not to be questioned, beyond doubt. (Years later, when I read *1984*, the climactic horror for me, the ultimate sign of Winston's disintegration and surrender, was his being forced, under torture, to deny that two and two is four. Even more terrible was the fact that eventually he began to doubt this in his own mind — that, finally, numbers failed him, too.)

I particularly loved prime numbers, the fact that they were indivisible, could not be broken down, were inalienably themselves. (I had no such confidence in myself, for I felt I was being divided, alienated, broken down more and more every week.) Why did primes come when they did? Was there any pattern, any logic to their distribution? Was there any limit to them, or did they go on forever? I spent innumerable hours factoring, searching for primes, memorizing them. They afforded me many hours of absorbed, solitary play, in which I needed no one else.

I made a grid, ten by ten, of the first hundred numbers, with the primes blacked in, but I could see no pattern, no logic to their distribution. I made larger tables, increased my grids to twenty squared, thirty squared, but still could discern no obvious pattern.

The only real holidays I had during the war were visits to a favorite aunt in Cheshire, in the midst of Delamere Forest, where she had founded the Jewish Fresh Air School for "delicate children." All the children, indeed, had little gardens of their own, squares of earth a couple of yards wide, bordered by stones. I wished desperately that I could go there, rather than Greystone — but this was a wish I never expressed (though I wondered if my clear-sighted and loving aunt did not divine it).

On my visits, Auntie Len always delighted me by showing me all sorts of botanical and mathematical pleasures. She showed me the spiral patterns on the faces of sunflowers in the garden, and suggested I count the florets in these. As I did so, she pointed out that they were arranged according to a series — 1, 1, 2, 3, 5, 8, 13, 21, etc. — each number being the sum of the two that preceded it. And if one divided each number by the number that followed it (1/2, 2/3, 3/5, 5/8, etc.), one approached the number 0.618. This series, she said, was called a Fibonacci series, after an Italian mathematician who had lived centuries before. The ratio of 0.618, she added, was known as the Divine Proportion, or Golden Section, an ideal geometrical proportion found in many plants and shells, and often used by architects.

She would take me for long, botanizing walks in the forest, where she had me look at fallen pinecones to see that they, too, had spirals based on the Golden Section. She showed me horsetails growing near a stream, had me feel their stiff, jointed stems, and suggested that I measure these when I got back to school and plot the lengths of the successive segments as a graph. When I did so, and saw that the curve flattened out, she explained that the increments were "exponential," and that this was the way growth usually occurred. These ratios, these geometric proportions, she told me, were to be found all over nature — numbers were the way the world is put together. Numbers, my aunt said, are the way God thinks.

The association of plants, of gardens, with numbers assumed a curiously intense, symbolic form for me. I started to think in terms of a kingdom or a realm of numbers, with its own geography, languages, and laws; but, even more, of a garden of numbers, a magical, secret, wonderful garden in which I could wander and play to my heart's content. It was a garden hidden from, inaccessible to, the bullies and the headmaster; and a garden, too, where I somehow felt welcomed and befriended. Among my friends in this garden were not only primes and Fibonacci sunflowers but perfect numbers (such as 6 or 28, the sum of their factors excluding themselves); Pythagorean numbers, whose square was the sum of two other squares (such as 3, 4, 5 or 5, 12, 13); and "amicable numbers" — pairs of numbers (such as 220 and 284) in which the factors of each added up to the other. And my aunt had shown me that my garden of numbers was doubly magical — not just delightful and friendly, always there, but part of the plan on which the whole universe was built.

III — Uncle Tungsten

I returned to London in the summer of 1943, after four years of exile, a ten-year-old boy, withdrawn and disturbed in some ways but with a passion for metals, for plants, and for numbers.

I delighted in being able to visit Uncle Tungsten again, and I think he also delighted in having his young protégé back, for he would spend hours with me in his factory and his lab, answering questions as fast as I could ask them.

He had several glass-fronted cabinets in his office, one of which contained a series of electric light bulbs: there were several Edison bulbs from the early 1880s with filaments of carbonized thread; a bulb from 1897, with a filament of osmium; and several bulbs from a few years later, with spidery filaments of tantalum tracing a zigzag course inside them.

Then there were the more recent bulbs — these were Uncle Dave's special pride and interest, for some he had pioneered himself — with tungsten filaments of all shapes and sizes. There was even one labeled "Bulb of the Future?" It had no filament, but the word *rhenium* was inscribed on a card beside it.

I had heard of platinum, but the other metals — osmium, tantalum, rhenium — were new to me. Uncle kept samples of them all, and some of their ores, in a cabinet next to the bulbs. As he handled them, he would expatiate on their unique, sovereign properties and qualities, how they had been discovered, how they were refined, and why they were so suitable for making filaments.

He would bring out a pitted gray nugget: "Dense, eh?" he would say, tossing it to me. "That's a platinum nugget. This is how it is found, as nuggets of pure metal. Most metals are found as compounds with other things, in ores. There are very few other metals which occur native like platinum — just gold, silver, and copper, and one or two others." These other native metals had been known, he said, for thousands of years, but platinum had been "discovered" only two hundred years ago, for though it had been prized by the Incas for centuries, it was unknown to the rest of the world. When the explorers brought it back, in the eighteenth century, the new metal enchanted all of Europe — it was denser, more ponderous than gold, and, like gold, it was "noble" and never tarnished. It had a luster exceeding that of silver. (Its Spanish name, *platina,* meant "little silver.")

Native platinum was often found with two other metals, iridium and osmium, which were even denser, harder, more refractory. Here Uncle pulled out samples for me to handle, mere flakes, no larger than lentils, but astoundingly heavy. These were "osmiridium," a natural alloy of osmium and iridium, the two densest substances in the world. There was something about heaviness, density — I could not say why — that gave me a thrill, and an immense sense of security and comfort. Osmium, moreover, had the highest melting point of all the platinum metals, so it was used at one time, Uncle said, to replace the platinum filaments in light bulbs.

The great virtue of the platinum metals was that they were as noble and workable as gold but had much higher melting points. Crucibles made of platinum could withstand the hottest temperatures; beakers and spatulas of it could withstand the most corrosive acids. Uncle Dave often used platinumware in his own lab, sometimes alloyed with other platinum metals to give it greater hardness and a still higher melting point. He pulled out a small crucible from the cabinet, beautifully smooth and shiny. It looked new. "This was made around 1840," he said. "A century of use, and almost no wear."

Uncle Dave saw the whole earth, I think, as a gigantic natural laboratory, where heat and pressure caused not only vast geologic movements but innumerable chemical miracles. "Look at these diamonds," he would say. "They were formed thousands of millions of years ago, deep in the earth, under unimaginable pressures."

He liked to pull out the native metals from his cabinet — twists and spangles of rosy copper, wiry silver, latticed gold. "Think how it must have been," he said, "seeing metal for the first time — sudden glints of reflected sunlight, sudden shinings in a rock or at the bottom of a stream!"

He would conjure up the first smelting of metal, how cavemen might have used rocks containing a copper mineral — green malachite, perhaps — to surround a cooking fire and suddenly realized as the wood turned to charcoal that the green rock was bleeding, turning into a red liquid: molten copper.

It took a much hotter fire, a white-hot fire, to obtain tungsten. Uncle Dave handed me a little ingot. "Tungsten," he said. "No one realized at first how perfect a metal tungsten was. It has the highest melting point of any metal, it is tougher than steel, and it keeps its strength at high temperatures — an ideal metal!"

Uncle had a variety of tungsten bars in his office — some he used as paperweights, but others had no discernible function whatever, except to give pleasure to their owner and maker. And, indeed, by comparison steel bars, and even lead, felt light and somehow porous, tenuous. "These lumps of tungsten have an extraordinary concentration of mass," he would say. "They would be deadly as weapons — far deadlier than lead."

But sooner or later Uncle's soliloquies and demonstrations before the cabinet all returned to tungsten's mineral ores. One of these, scheelite, was named after the great Swedish chemist Carl Wilhelm Scheele, who was the first to show that it contained a new element. The ore was so dense that miners called it "heavy stone," or *"tung-sten,"* the name subsequently given to the element itself. Scheelite was found in beautiful orange crystals that fluoresced bright blue in ultraviolet light. Uncle Dave kept specimens of scheelite and other fluorescent minerals in a special cabinet in his office. The dim light of Farringdon Road on a November evening, it seemed to me, would be transformed when he turned on his Wood's lamp, and the luminous chunks in the cabinet suddenly glowed orange, turquoise, crimson, green.

Though scheelite was the largest source of tungsten metal, the metal had first been obtained from a different mineral, called wolframite. Indeed, tungsten was sometimes called wolfram, and still retained the chemical symbol *W.* This thrilled me, because my own middle name was Wolf. Heavy seams of the tungsten ores were often found along with tin ore, and the tungsten made it more difficult to isolate the tin. This was why, my uncle continued, they had originally called the metal wolfram — for, like a hungry animal, it "stole" the tin. I liked the name wolfram, its sharp, animal quality, its evocation of a ravening, mystical wolf — and thought of it as a tie between Uncle Tungsten, Uncle Wolfram, and myself, O. Wolf Sacks.

Names fascinated me — their sounds, their associations, the sense they gave of people and places. The names of the elements were filled with such evocations, but there were only a few dozen of these, whereas the number of minerals ran into the hundreds or thousands. These were all beautifully laid out, with their names and formulas, in the cabinets of the Geology Museum, in South Kensington, where, later, I would go whenever I could.

The older names gave one a sense of antiquity and alchemy: corundum and galena, orpiment and realgar. (Orpiment and realgar, two arsenic sulfides, went euphoniously together and made me think of an operatic couple, like Tristan and Isolde.) There was pyrite, fool's gold, in brassy, metallic cubes, and chalcedony and ruby and sapphire and spinel. Zircon sounded oriental; calomel, Greek — its honeylike sweetness, its "mel," belied by its poisonness. There was the medieval-sounding sal ammoniac. There was cinnabar, the heavy red sulfide of mercury, and massicot and minimum, the twin oxides of lead.

Then there were minerals named after people. One of the most common minerals, much of the redness of the world, was the hydrated iron oxide called goethite. (Was this named in honor of Goethe, or did he discover it? I had read that he had a passion for mineralogy and chemistry.) Many minerals were named after chemists — gaylussite, scheelite, berzelianite, bunsenite, liebigite, moissanite, crookesite, and the beautiful, prismatic "ruby-silver," proustite. There was samarskite, named after a mining engineer, Colonel Samarski. There were other names that were evocative in a more topical way: stolzite, a lead tungstate, and scholzite, too. Who were Stolz and Scholz? Their names seemed very Prussian to me, and this, just after the war, evoked an anti-German feeling. I imagined Stolz and Scholz as Nazi officers with barking voices, swordsticks, and monocles.

Other names appealed to me mostly for their sound, and for the images they conjured up. I loved classical words and their depiction of simple properties — the crystal forms, colors, shapes, and optics of minerals — like diaspore and anastase and microlite and polycrase. A great favorite was cryolite — ice stone, from Greenland, so low in refractive index that it was transparent, almost ghostly, and, like ice, became invisible in water.

On one visit, Uncle Dave showed me a large bar of aluminum. After the dense platinum metals, I was amazed at how light it was, scarcely heavier than a piece of wood. "I'll show you something interesting," he said. He took a smaller lump of aluminum, with a smooth, shiny surface, and smeared it with mercury. All of a sudden — it was like some terrible disease — the surface broke down, and a white substance like a fungus rapidly grew out of it, until it was a quarter of an inch high, then half an inch high, and it kept growing and growing until the small piece of aluminum was completely eaten up. "You've seen iron rust, oxidizing, combining with the oxygen in the air," Uncle said. "But here, with the aluminum, it's a million times faster. That big bar is still quite shiny, because it's covered by a fine layer of oxide, and that protects it from further change. But rubbing it with mercury destroys the surface layer, so then the aluminum has no protection, and it combines with the oxygen in seconds."

I found this magical, astounding, but also a little frightening — to see a bright and shiny metal reduced so quickly to a crumbling mass of oxide. It made me think of a curse or a spell, the sort of disintegration I sometimes saw in my dreams. It made me think of mercury as evil, as a destroyer of metals. Would it do this to every sort of metal?

"Don't worry," Uncle answered. "The metals we use here, they're perfectly safe. If I put this little bar of tungsten in the mercury, it would not be affected at all. If I put it away for a million years, it would be just as bright and shiny as it is now." In a precarious world, tungsten, at least, was stable.

As the youngest of almost the youngest (I was the last of four, and my mother the sixteenth of eighteen), I was born nearly a hundred years after my maternal grandfather, and never knew him. He was born Mordechai Fredkin, in 1837, in a small village in Russia. As a youth, he managed to avoid being impressed into the Cossack army and fled Russia, using the passport of a dead man named Landau. He was sixteen. As Marcus Landau, he made his way to Paris, and then Frankfurt, where he married. (His wife was sixteen, too.) Two years later, in 1855, now with the first of their children, they moved to England.

My mother's father was, by all accounts, a man drawn equally to the spiritual and the physical. He was by profession a boot and shoe manufacturer, a *shochet* (a kosher slaughterer), and later a grocer — but he was also a Hebrew scholar, a mystic, an amateur mathematician, and an inventor. He had a wide-ranging mind: he published a newspaper, the *Jewish Standard,* in his basement, from 1888 to 1891; he was interested in the new science of aeronautics and corresponded with the Wright brothers, who paid him a visit when they came to London in the early 1900s. (Some of my uncles could still remember this.) He had a passion, my aunts and uncles told me, for intricate arithmetical calculations, which he would do in his head while lying in the bath. But he was drawn, above all, to the invention of lamps — safety lamps for mines, carriage lamps, street lamps — and he patented many of these in the 1870s. When I was very small, my mother would take me to the Science Museum, in South Kensington, up to the top floor, where there was a simulacrum of a coal mine, its dim, low passages lit by fitful beams. There she would show me the Landau safety lamp on display, right next to the more famous Humphry Davy lamp.

A polymath and an autodidact himself, Grandfather was passionately keen on education — and, most especially, a scientific education — for all his children, for his nine daughters no less than for his nine sons. Whether it was this or the sharing of his own passionate enthusiasms, seven of his sons were eventually drawn to mathematics and the physical sciences — including the two I was closest to, Uncle Dave and Uncle Abe.

IV — Stinks and Bangs

My parents and my brothers had introduced me, even before the war, to some kitchen chemistry: pouring vinegar on a piece of chalk in a tumbler and watching it fizz; then pouring the heavy gas this produced, like an invisible cataract, over a candle flame, putting it out straightaway. Or taking red cabbage, pickled with vinegar, and adding household ammonia to neutralize it. This would lead to an amazing transformation, the juice going through all sorts of colors, from red to various shades of purple, to turquoise and blue, and finally to green. I enjoyed these experiments, I wondered what was going on, but I did not feel a real chemical passion — a desire to compound, to isolate, to decompose, to see substances changing, familiar ones disappearing and new ones appearing in their stead — until I returned from Greystone, remet Uncle Dave, and saw his lab and his passion for experiments of all kinds.

Now, after hearing him talk about chemistry, and starting to read about chemistry and chemists myself, I longed to have a lab of my own.

As a start, I wanted to lay hands on cobaltite and niccolite, and compounds or minerals of manganese and molybdenum, of uranium and chromium — all those wonderful elements that were discovered in the eighteenth century. I wanted to pulverize them, treat them with acid, roast them, reduce them — whatever was necessary — so I could extract their metals myself. I knew, from looking through a chemical catalogue at the factory, that one could buy these metals already purified, but it would be far more fun, far more exciting, I reckoned, if I was able to make them myself. This way, I would enter chemistry, start to discover it for myself, in much the same way its first practitioners did — I would live the history of chemistry in myself.

It was through reading Mary Elvira Weeks's *Discovery of the Elements* — a book published just before the war — that I got a vivid idea of the lives of many chemists, the great variety, and sometimes vagaries, of character they showed, and the relation (sometimes) between their characters and their work. Here I found quotations from the early chemists' letters, which portrayed their excitements (and despairs) as they fumbled and groped their way to their discoveries, losing the track now and again, getting caught in blind alleys, though ultimately reaching the goal they sought.

If Humphry Davy was the first chemist I had ever heard of, he was also the one I most warmed to. I loved reading of his experiments with explosives and electric fish; his discovery of incandescent metal filaments and electric arcs; of catalysts (it was only now that I realized why we had a platinum loop above the gas stove); of the physiological effects of nitrous oxide, laughing gas; and, above all, of his using the just invented electric battery to isolate the alkali and alkaline-earth metals in a single miraculous year. He appealed to me especially because he was boyish and impulsive, the way he danced with joy all around his lab when he first isolated potassium, in 1807, and saw the shining metallic globules burst and take fire. Davy moved me to emulation — to sampling the effects of nitrous oxide for myself (my mother kept a cylinder of it in her obstetric bag), and making my own sodium and potassium by electrolysis.

I was awed, too, by the figure of Mendeleev — his passionate search for order among the elements (more than sixty were known by the 1850s, a rich chaos), and his final discovery of such an order (supposedly in a dream) in 1869. When I first saw the periodic table, it hit me with the force of revelation — it embodied, I was convinced, eternal truths, the eternal and necessary order of the elements. I thought of Mendeleev as a sort of Moses, bearing the tablets of the God-given Periodic Law.

And then, in a different mode, there was Marie Curie, who

had spent four backbreaking years in a shed extracting a pinch of "her" element, radium, from four stubborn tons of pitchblende. Radium, my mother would say, was a magical element, with unique powers to harm and to cure. She herself had worked with radium therapy at the Marie Curie Hospital, in London, and had met Marie Curie on one occasion. (I was intrigued when she told me of the radium "bomb" at the hospital, and the fine gold needles full of radon that were inserted into tumors.) Eve Curie's biography of her mother — which my mother gave me when I was ten — was the first portrait of a scientist I read, and Marie Curie was added to my pantheon of heroes. (Fifty-five years later, in 1998, at a meeting in New York to celebrate the centenary of the Curies' discovery of polonium and radium, I met Eve Curie, now in her nineties, and asked her to sign the book.)

It was through reading these accounts that I first realized one could have heroes in real life. There seemed to me an integrity, an essential goodness, about a life dedicated to science. I had never given much thought to what I might be when I was "grown up" — growing up was hardly imaginable — but now I knew: I wanted to be a chemist. A sort of eighteenth-century chemist coming fresh to the field, looking at the whole, undiscovered world of natural substances and minerals, analyzing them, plumbing their secrets, finding the wonder of new and unknown metals.

And so I set up a little lab of my own at home. There was an unused back room I took over, originally a laundry room, which had running water and a sink and drain and various cupboards and shelves. Conveniently, this room led out to the garden — so that if I concocted something that caught fire, or boiled over, or emitted noxious fumes, I could rush outside with it and fling it on the lawn. The lawn soon developed charred and discolored patches, but this, my parents felt, was a small price to pay for my safety — their own too, perhaps. But seeing occasional flaming globules rushing through the air, and the general turbulence and abandon with which I did things, they were alarmed, and urged me to plan experiments and to be prepared to deal with fires and explosions. Eventually, after I had filled the house one day with vile-smelling (and very toxic) hydrogen sulfide, they insisted that I install a small fume cupboard and a special drain for corrosive liquids — and that with dangerous experiments I wear gloves and goggles.

Uncle Dave advised me closely on the choice of apparatus — test tubes, flasks, graduated cylinders, funnels, pipettes and burettes, a Bunsen burner, crucibles, watch glasses, a platinum loop, a desiccator, a blowpipe, a range of spatulas, a balance. He advised me, too, on basic reagents, some of which he gave me from his own lab, along with a supply of stoppered bottles of all sizes — bottles of varied shapes and colors (dark green or brown for light-sensitive chemicals), with perfectly fitting ground-glass stoppers.

Every month or so, I stocked my lab with visits to a chemical-supply shop, Griffin & Tatlock, far out in Finchley. The shop was housed in a large shed set at a distance from its neighbors (who viewed it, I imagined, with a certain trepidation, as a place that might explode or exhale poisonous fumes at any moment). I would hoard my pocket money for weeks — occasionally one of my uncles, approving my secret passion, would slip me a half-crown or so — and then take a succession of trains and buses to the shop. The cheaper chemicals were kept in huge stoppered urns of glass. The rarer, more costly substances were kept in smaller bottles. Hydrofluoric acid — dangerous stuff, used for etching glass — could not be kept in glass, so it was sold in special small bottles made of gutta-percha, a sort of rubbery substance. Beneath the serried urns and bottles on the shelves were great carboys of acid — sulfuric, nitric, aqua regia; globular china bottles of mercury (it was incredibly dense, and seven pounds of it would fit into a bottle the size of a fist); and slabs and ingots of the commoner metals. The shopkeepers soon got to know me — an intense and rather undersized schoolboy, clutching his pocket money, spending hours amid the jars and bottles — and though they would warn me now and then, "Go easy with that one!" they always let me have what I wished.

My first taste was for the spectacular — the frothings, the incandescences, the stinks and the bangs, which almost define an entry into chemistry. One of my guides was J. J. Griffin's *Chemical Recreations,* an 1850-ish book I had found in a secondhand bookshop. Griffin started recreational and gradually got more systematic. I worked my way through "Alteration of Vegetable Colours by Acids and Alkalis," "Experiments with Coloured Liquors and Sympathetic Inks," "Chemical Metamorphoses," and then got on to the serious stuff. There was a special chapter on "Chemistry for Holidays," with the "Volatile Plum-Pudding" ("when the cover is removed . . . it leaves its dish and rises to the ceiling"), "A Fountain of Fire" (using phosphorus — "the operator must take care not to burn himself"), and "Brilliant Deflagration" (here, too, one was warned to "remove your hand instantly"). I was amused by the mention of a special formula (sodium tungstate) to render ladies' dresses and curtains incombustible — were fires that common in Victorian times? — and used it to fireproof a handkerchief for myself.

Chemical exploration, chemical discovery, if full of excitement, was full of surprises and dangers, too. I was struck by the range of accidents that had befallen the pioneers. Few naturalists had been devoured by wild animals or stung to death by noxious plants or insects; few physicists had lost their eyesight looking at the heavens, or broken a leg on an inclined plane; but chemists had lost their eyes, their limbs, and sometimes their lives, usually by causing explosions or producing inadvertent toxins. Davy, for instance, had been nearly asphyxiated by nitric oxide, had poisoned himself with nitrogen peroxide, and had severely inflamed his lungs with hydrofluoric acid, prior to his dangerous experiments with nitrogen trichloride.

Bunsen, investigating cacodyl cyanide, lost his right eye in an explosion, and very nearly lost his life. Several experimenters, trying to make diamond from graphite in intensely heated, high-pressure "bombs," threatened to blow themselves and their fellow workers to kingdom come.

Mary Elvira Weeks's book on the discovery of the elements devoted an entire section to "The Fluorine Martyrs." All the early experimenters, I read, had "suffered the frightful torture of hydrofluoric-acid poisoning," and at least two of them died in the process. After reading this, I was too scared to open the hydrofluoric acid I had bought.

Attracted by the sounds and flashes and smells coming from my lab, my two older brothers, Marcus and David, now medical students, sometimes joined me in experiments — the ten-year age difference between us hardly mattered at these times. On one occasion, as I was experimenting with hydrogen and oxygen, there was a loud explosion, and an almost invisible sheet of flame, which blew off Marcus's eyebrows completely. But Marcus took this in good part, and he and David often suggested other experiments.

We mixed potassium perchlorate with sugar, put it on the back step, and banged it with a hammer. This caused a most satisfying explosion.

We made a "volcano" together with ammonium dichromate, setting fire to a pyramid of the orange crystals, which then flamed, furiously, becoming red-hot, throwing off showers of sparks in all directions, and swelling portentously, like a miniature volcano. Finally, when it had died down, there was, in place of the neat pyramid of crystals, a huge, fluffy pile of dark-green chromic oxide.

Another early experiment, suggested by David, was pouring oily, concentrated sulfuric acid on a little sugar, which instantly turned black, heated, steamed, expanded, forming a monstrous pillar of carbon that rose high above the rim of the beaker. "Beware," David said as I gazed at this transformation. "You'll be turned into a pillar of carbon if you get the acid on yourself."

The first recorded individual who discovered an element, it seems, was Hennig Brandt, of Hamburg, who obtained phosphorus (apparently with some alchemical ambition in mind) from urine in 1669. He adored the strange, luminous element and called it "cold fire" (*kaltes Feuer*), or, in a more affectionate mood, *"mein Feuer."* Throughout the eighteenth century it was made from bones.

I decided to obtain my phosphorus direct from Griffin & Tatlock. When it came, as a bundle of pale, waxy sticks that one had to keep under water, it had, nonetheless, a persistent garlicky smell — and this, I imagined, was the irremovable residue of its beastly, slaughterhouse origins. It was important to keep it in its brown bottle, because the beautiful, almost colorless translucent sticks became ugly and yellow and opaque with the light — another example of the mysterious power of light.

Phosphorus attracted me strangely, dangerously, because of its luminosity — I would sometimes slip down to my lab at night to gaze at it — the "cold fire" that had so fascinated its discoverer (and had caused him, and others, such terrible burns). A whole series of experiments, of enchantments, spread out from this. As soon as I had my fume cupboard set up, I put a piece of white phosphorus in water and boiled it, dimming the lights so that I could see the steam coming out of the flask, glowing a soft greenish blue.

If one ignited phosphorus in a bell jar (using a magnifying glass), the jar would fill with a "snow" of phosphorus pentoxide. If one did this over water, the pentoxide would hiss, like red-hot iron, as it hit the water and dissolved, making phosphoric acid.

Another, rather beautiful, experiment was boiling phosphorus with caustic potash in a retort — I was remarkably nonchalant in boiling up such virulent substances; lucky, too, in that I never really hurt myself — and this produced phosphoretted hydrogen (the old term), or phosphine. As the bubbles of phosphine escaped, they took fire spontaneously, forming beautiful rings of white smoke.

By heating white phosphorus, I could transform it into a much stabler form — red phosphorus, the phosphorus of matchboxes. I had learned as a small child that diamond and graphite were different forms, allotropes, of the same element, and I vividly remembered how my shining tin soldiers had been transformed one winter into a gray dust: "tin pest." Now, in the lab, I could effect some of these changes for myself, turning white phosphorus into red phosphorus, and then (by condensing its vapor) back again. These transformations made me feel like a magician.

While I darted about these exotic experiments, I also went steadily through my small repertoire of chemicals, trying to learn their basic properties and reactions, and the basics of forming and decomposing compounds.

I decomposed water using a large battery, much as Humphry Davy had done at the start of the nineteenth century; and then I recomposed it, sparking hydrogen and oxygen together. There were many other ways of making hydrogen with acids or alkalis — with zinc and sulfuric acid or aluminum bottle caps and caustic soda.

But it seemed a shame to have the hydrogen just bubble off and go to waste. To stopper my flasks, I got some tight-fitting rubber bungs and corks, some with holes in the middle for glass tubes. One of the things I had learned in Uncle Dave's lab was how to soften glass tubing in a gas flame, and gently bend it to an angle (and, more exciting, to blow glass as well, gently puffing into the molten glass to make thin-walled globes and shapes of all sorts). Now, using glass tubing, I could light the hydrogen as it emerged from the stoppered flask. It had a colorless flame — not yellow and smoky like the flames of gas jets or the kitchen stove. Or I could feed the hydrogen, with a gracefully curved piece of glass

tubing, into a soap solution, to make soap bubbles filled with hydrogen; the bubbles, far lighter than air, would rush up to the ceiling and burst.

I made oxygen, too. I wanted to make it by heating mercuric oxide — this was the original way, I read, by which Joseph Priestley
had first made it, in 1774 — but I was afraid, until the fume cupboard was installed, of toxic mercury fumes. Yet it was easy to prepare by heating an oxygen-rich substance like deep purple-red potassium permanganate. I remember thrusting a glowing wood chip
into a test tube full of oxygen, and seeing how it flared up, flamed,
with an intense brilliance.

There were some metals that were so reactive they could actually
tear the oxygen out of water, leaving the hydrogen to bubble off.
Magnesium did this if the water was hot — this was why one could
not put out an incendiary bomb with water. And potassium did so
explosively, even in cold water.

Sodium was cheaper, and not quite as violent as potassium, so I
decided to look at its action outdoors. I obtained a good-sized
lump of it — about three pounds — and made an excursion to
Highgate Pond, in Hampstead Heath, with two of my friends, Eric
Korn and Jonathan Miller. When we arrived, we climbed up a little
bridge, and then I pulled the sodium out of its oil with tongs and
flung it into the water beneath. It took fire instantly and sped
around and around on the surface like a demented meteor, with a
huge sheet of yellow flame above it. My friends still have vivid memories of this. We all exulted — this was chemistry with a vengeance!

V — Brilliant Light

This first interest in chemistry, this desire to interrogate and explore everything in sight, led to a different feeling after a couple of
years — a need to integrate my knowledge, to understand. Not just
to throw sodium into water but all the alkali metals, to see how they
compared, to plot the trends, physical and chemical, as one went
through them all, from lithium to sodium to potassium to rubid-
ium to cesium. Weighing and measuring became crucially important, convincing me of the fixed proportions in which elements
combined, drawing me to atomic theory and the concept of atomic
weights.

Reading Dalton, who proposed the atomic hypothesis, reading

about his atoms (I saw his own wooden models of these in the science museum), put me in a sort of rapture, conceiving that the chemical reactions one saw macroscopically in the lab, with all their puzzling constancies and exactitudes, were the consequence of activities almost infinitely small — single atoms, with distinctive weights and characters, combining with each other one by one — and imagining that if one had a microscope powerful enough, and powers of perception quick enough, this was what one would actually see. Until now, there had been only a vague, mysterious sense of an invisible microworld; Dalton's atoms gave the imagination something concrete to chew on, made this tangible and real.

It was only when I had the concept of atomic weight firmly in mind, along with the concept of atoms' combining power, or valency, that I could appreciate the startling beauty, the obvious truth, of the periodic table — for me, now, the most beautiful thing in the world. By arranging all the elements in a simple grid of intersecting "periods" and "groups," the table showed, at a glance, how all of them were related to one another, and how one could predict the existence and properties of as yet undiscovered elements simply by knowing their place in the table, for when one arranged the elements in order of atomic weight their properties echoed one another periodically. Thus, each period started with an alkali metal and ended with an inert gas, and then one shuttled back to the next period, starting with a heavier alkali metal and ending with a heavier inert gas. The periods contained eight elements apiece at first, then expanded to eighteen, then thirty-two — a mysterious but surely fundamental numerical series. I could not help thinking back to the grids, the tables of primes, I used to make, where I sought so desperately for order but found none. The periodic table, by contrast, was a Jacob's ladder, a numinous spiral, going up to, coming down from, a Pythagorean heaven.

This almost religious feeling about the periodic table afforded me a very deep, cosmic sense of security and stability, a conviction that the physical universe, at least, was lawful, orderly, harmonious. This certainly did much to alleviate the terrifying uncertainties that had so undone me in my years at Greystone.

Besides spending time with Uncle Dave and in my lab, I began to spend time with another of my mother's brothers, Uncle Abe. He was a few years older than Dave, and more disposed to physics than to chemistry: the great discoveries of X-rays, radioactivity, the electron, and quantum theory had all occurred in his formative years.

Though Abe and Dave were alike in some ways (both had the broad Landau face, with wide-set eyes, and the unmistakable, resonant Landau voice — characteristics still marked in the great-great-grandchildren of my grandfather), they were very different in others. Dave was tall and strong, with a military posture (he had served in the Great War, and in the Boer War before that), always carefully dressed. He would wear a wing collar and highly polished shoes even when he worked at his lab bench. Abe was smaller, somewhat gnarled and bent (in the years that I knew him), brown and grizzled, like an old shikari, with a hoarse voice and chronic cough; he cared little what he wore, and usually had on a sort of rumpled lab smock.

Both Uncle Dave and Uncle Abe were intensely interested in light and lighting, but with Dave it was "hot" light, and with Abe "cold" light. Uncle Dave had drawn me into the history of incandescence, of the rare earths and metallic filaments that glowed and incandesced brilliantly when heated. He had inducted me into the energetics of chemical reactions — how heat was absorbed or emitted during the course of these; heat that sometimes became visible as fire and flame.

Through Uncle Abe, I was drawn into the history of "cold" light — luminescence — which started perhaps before there was any language to record things, with observations of fireflies and glowworms and will-o'-the-wisps and luminous seas, and of Saint Elmo's fire, the eerie luminous discharges that could stream from a ship's masts, giving its sailors a feeling of bewitchment.

Abe's first love was the investigation of fluorescence and phosphorescence — the power of certain substances to absorb radiant energy like ultraviolet or X-rays and reemit it as light in the visible range — and he had developed and patented a luminous paint containing radium that was used in military gunsights during the First World War (and may have been decisive, he told me, in the Battle of Jutland).

His attic was a wondrous place, full of electric machines and induction coils, batteries and magnets, and sealed vacuum tubes (Geissler tubes) of rarefied gases, which when electrified would light up with brilliant colors — neon red, helium yellow. It was here, with Abe, that I learned about electricity and color and fluorescence; with Abe (and his ten milligrams of radium bromide) that I learned something about the wonders and dangers of radioactivity. (His own hands were covered with radium burns and malignant warts, from his long, careless handling of radioactive materials.) It was with Abe, too, that I learned about spectroscopy — putting different elements into the colorless flame of a Bunsen burner and looking at the radiant spectra they emitted; seeing how these were unique to each element. Spectroscopy became one of my delights as a boy. Most especially, Abe dwelt on the mysterious spacing of the spectral lines, how (at least in hydrogen) they followed a simple formula, and how this was to remain tantalizingly unexplained for almost thirty years, like the periodicity of the elements in the periodic table itself.

It was through Uncle Abe that I learned of the incredible scientific events of 1913. It was "the year the world changed," he would always say, with Bohr's quantal model of the atom and Moseley's X-ray spectrography of the elements confirming between them the periodic table, providing an electronic understanding of the chemical properties of the elements (as well as the spacing of their spectral lines), in terms of a new and radical theory of the atom.

Prior to this, as Bohr remarked, spectra seemed as beautiful and meaningless as butterflies' wings, but now, in the words of another great pioneer, Arnold Sommerfeld, they were revealed as "a true atomic music of the spheres." Uncle Abe spoke of the sense of brilliant light, of sudden, profound understanding, that came to many chemists and physicists at this time, and how it was suddenly overwhelmed by the terribleness of the Great War.

VI — The End of the Affair

With the ending of the war, other triumphs were soon to come: an understanding of why metals were lustrous, why they conducted heat and electricity, the nature of color, of luminescence, of density, of magnetism — all the questions I had puzzled over as a boy. There was an exuberance at this. But the promise held a threat, too: What would become of classical chemistry now? What need was there for it, if the new theory was so powerful?

I had dreamed of becoming a chemist, but the chemistry that really stirred me was the lovingly detailed, naturalistic, descriptive chemistry of the nineteenth century, not the new chemistry of the quantum age, which, so far as I understood it, was highly abstract and, in a sense, closer to physics than to chemistry. Chemistry as I knew it, the chemistry I loved, was either finished or changing its character, advancing beyond me.

From this point, chemistry seemed to recede from my mind — my love affair, my passion for it, came to an end. There were, perhaps, other factors as well. I had been living (it seems to me in retrospect) in a sort of sweet interlude, having left behind the horrors and fears of Greystone. I had been guided to a region of order, and a passion for science, by two very wise, affectionate, and understanding uncles. My parents had been supportive and trusting, had allowed me to put a lab together and follow my own whims. School, mercifully, had been indifferent to what I was doing — I did my schoolwork, and was otherwise left to my own devices. Perhaps, too, there was a biological respite, the special calm of latency.

But now all this was to change: adolescence rushed upon me, like a typhoon, buffeting me with insatiable longing. And at school I left the undemanding classics "side" and moved to the pressured science side instead. I had been spoiled, in a sense, by my two uncles, and the freedom and spontaneity of my apprenticeship. Now, at school, I was forced to sit in classes, to take notes and exams, to use textbooks that were flat, impersonal, deadly. What had been fun, delight, when I did it in my own way became an aversion, an ordeal, when I had to do it to order. In his essay "To Teach or Not to Teach," Freeman Dyson speaks of different sorts of people: students who are best given independence, allowed and encouraged to develop in their own way, and those who profit most from structured teaching, from school. I was clearly one who flourished best alone.

Was it, then, the end of chemistry? My own intellectual limitations? Adolescence? School? Or was it, more simply, that I was growing up, and that "growing up" makes one forget the lyrical, mystical perceptions of childhood, the glory and the freshness of which Wordsworth wrote, so that they fade into the light of common day?

*

After all, it was "understood," by the time I was fourteen, that I was going to be a doctor; my parents were doctors, my brothers in medical school. My parents had been tolerant, even pleased, with my early interests in science, but now, they seemed to feel, the time for play was over. One incident stays clearly in my mind. In 1947, a couple of summers after the war, I was with my parents in our old Humber touring the South of France. Sitting in the back, I was talking about thallium, rattling on and on about it: how it was discovered, along with indium, in the 1860s by the brilliant-colored green line in its spectrum; how some of its salts, when dissolved, could form solutions nearly five times as dense as water; how thallium indeed was the platypus of the elements, with paradoxical qualities that had caused uncertainty about its proper placement in the periodic table — soft, heavy, and fusible, like lead, chemically akin to gallium and indium, but with dark oxides like those of manganese and iron, and colorless sulfates like those of sodium and potassium. As I babbled on, gaily, narcissistically, blindly, I did not notice that my parents, in the front seat, had fallen completely silent, their faces bored, tight, and disapproving — until, after twenty minutes, they could bear it no longer, and my father burst out violently: "Enough about thallium!"

No doubt anyone would have responded the same way. But now, the message came through, it was time to grow up, to be serious, to work — these words were used again and again — for the training of a doctor was long, hard, and demanding.

The old lab bench has now become a thing of the past. When I paid a visit not long ago to the old building in Finchley that had been Griffin & Tatlock's home a half century ago, it was no longer there. Such shops, such suppliers, which had provided chemicals and simple apparatus and unimaginable delights for generations, have now all but vanished.

And yet the old enthusiasm, which I had thought dead at fourteen, survives, clearly, deep inside me, and surfaces every so often in odd associations and impulses. A sudden desire for a ball of cadmium, or to feel the coldness of diamond against my face. The license plates of cars immediately suggest elements, especially in New York, where so many of them begin with U, V, W, or Y — that

is, uranium, vanadium, tungsten, and yttrium. It is an added pleasure, a bonus, a grace, if the symbol of an element is followed by
its atomic number, as in W 74 or Y 39. Flowers, too, bring elements to mind: the color of lilacs in spring for me is that of divalent
vanadium.

I often dream of chemistry at night, dreams that conflate the
past and the present, the grid of the periodic table transformed to
the grid of Manhattan. The location of tungsten, at the intersection of Group VI and Period 6, becomes synonymous here with the
intersection of Sixth Avenue and Sixth Street. (There is no such intersection in New York, of course, but it exists, is conspicuous, in
the New York of my dreams.) I dream of eating hamburgers made
of scandium. Sometimes, too, I dream of the indecipherable language of tin (a confused memory, perhaps, of its plaintive "cry").
But my favorite dream is of going to the opera (I am Hafnium),
sharing a box at the Met with the other heavy transition metals —
my old and valued friends — Tantalum, Tungsten, Rhenium, Osmium, Iridium, Platinum, and Gold.

Yet it is not just dreams and associations, nostalgic yearnings,
that prick my imagination now but hearing of the new achievements of chemistry — a chemistry that, if less personal than the
old, has come to embrace the whole world and the universe. In my
day, elements stopped with No. 92, uranium, but now elements
up to 118 have been made. These new elements exist only in the
lab, and may not occur anywhere else in the universe. The very latest of them, though still radioactive, belong to a long-sought "island of stability," where their atomic nuclei are almost a million
times stabler than those of the preceding elements. We have seen
moon rocks and Mars rocks containing minerals never before seen.
We wonder about giant planets with cores of metallic hydrogen,
stars made of diamond, and stars with crusts of iron helide. The inert gases have been coaxed into combination, and I have seen a
fluoride of xenon, unthinkable in the 1940s. A totally unexpected
new form of carbon has been made, forming exquisite, soccerball-
shaped giant molecules (buckminsterfullerene, "bucky-balls" for
short). Scandium is now used in the fins of missiles — and in bicy-
cles. The rare-earth elements, which both Uncle Tungsten and Uncle Abe so loved, have now become common and found countless
uses in fluorescent materials, phosphors of every color, high-tem-

perature superconductors, and tiny magnets of an unbelievable strength.

These things, and a thousand others, excite me, stir me, set the imagination running in every direction; and they show me that, though I and the world may have changed beyond recognition, the boy who loved chemistry is still kicking inside.

HAMPTON SIDES

This Is Not the Place

FROM *DoubleTake*

NEAR THE TOWN OF PALMYRA, NEW YORK, rising over corn-
fields and dairy farms and the dark green thread of the Erie Canal,
is a glacially formed monadnock known as the Hill Cumorah. It's
too small to qualify as a mountain, but in its context Cumorah is an
arresting sight, wildly out of scale with the somnolent farm country
of New York's Finger Lakes region, like an interloper from a distant
geological epoch. At the hill's summit is an American flag, an asphalt pathway lined with pink rosebushes, and a golden statue of
the angel Moroni, from the Book of Mormon.

I had come to this distinctive landmark one muggy evening in
mid-July to watch the largest outdoor play in America, the Hill
Cumorah Pageant, a two-hour spectacle that features a cast of over
six hundred people. It's a kind of passion play that's been held in a
grass field at the base of the hill every July for sixty-one years. When
I arrived, an immense proscenium had been erected, and orches-
tral music was pouring through concert speakers. A crowd estimated at slightly more than ten thousand people had turned out
for this, the seventh and final performance of the 1998 pageant.
Along the edges of the field, hundreds of families were splayed
out on blankets enjoying the cool air of twilight. Ruritans were selling hamburgers and personal pizzas, and cast members in bibli-
cal attire — deerskin robes, leather sandals, and long false beards
— were ushering late arrivals to the last empty rows of plastic seats
in the rear. Then the sun went down, and in a blaze of trumpets
and laser lights swirling through smoke, the 627 actors gathered
on the stage.

The Hill Cumorah Pageant tells the tale, in a drastically distilled form, of the Book of Mormon. The play traces the family history of the Nephites, a tribe of Jews who leave Jerusalem around 600 B.C., journey on foot across the desert, and then set sail for a promised land. They faithfully drift across the ocean, *Kon-Tiki* fashion, and, after many disasters at sea, come to light somewhere on American shores. Once established in the New World, the Nephites build impressive cities of stone and do remarkable work with agriculture and metallurgy, when they're not battling their chief adversaries, a crude band of Indians called the Lamanites, who wear antlers and feathered headdresses and look vaguely like the Aztecs. Christ makes a brief appearance in America, and there are wilderness wanderings, cataclysmic storms, even a volcanic eruption, with plumes of steam and potato flakes to simulate ash. The story ends with a great battle on the Hill Cumorah in which the Nephites are finally exterminated by the Lamanites. After the dust settles, only one Nephite remains — Moroni, son of the supreme commander, Mormon. It is Moroni's solemn duty to take the ancient records, engraved on a set of golden plates, and bury them in the hill so that someone, one day, will learn the true story of America's lost tribe of Hebrews.

As a coda to the play, the story jumps forward some 1,400 years to 1823. The spotlights are trained on a young man climbing high along the west face of the Hill Cumorah, while celestial strains of the Mormon Tabernacle Choir seep from the concert speakers. He kneels while the angel Moroni points to the spot where the golden plates are buried. The young man is the prophet Joseph Smith, and the record he removes from this hallowed ground is the Book of Mormon.

After the pageant I met a cast member, Sister Spencer from Michigan, a vivacious woman in her mid-forties who was stationed in a semiofficial capacity at the base of the statue of Moroni to answer any questions people might have about the import of the play.

"Whatever happened to the golden plates?" I asked her. "Are they in a museum somewhere?"

"No, they were returned to the angel Moroni, probably reburied somewhere," Sister Spencer said. "There are individuals in the church who would like to find them. But God will reveal them only when and if he wants to."

"Where did all of these events take place?" I asked her. "The wars, the civilizations?"

"Well," Sister Spencer said, "Joseph found the plates here, we know that. But we're not sure about the rest of it. Some of the scholars are now saying it all happened in southern Mexico."

"In Mexico?" I asked.

"That's what the experts at BYU are saying — Mexico, Central America. The Mayans and all those people down there. Those wonderful ruins."

This geographical leap seemed to me an implausible new wrinkle in an already implausible saga, but Sister Spencer's statement about the scholars at Brigham Young University, I would discover, was correct. While church leaders in Salt Lake City have made no official pronouncements on the subject, the prevailing view within Mormon intellectual circles is that the primary action in the Book of Mormon did not, in fact, happen in upstate New York, but in Mesoamerica. During the past half century, the Church of Jesus Christ of Latter-Day Saints has been quietly attempting to prove this new theory. Over the years, the church and wealthy Mormon benefactors have sunk what is conservatively estimated to be $10 million into archaeological research all across Central America in what may be the most ambitious hunt for a vanished civilization since Schliemann's search for Troy.

Much of the excavation work has been the stuff of scrupulous scholarship carried out under the auspices of the Mormon-funded New World Archaeological Foundation, based in San Cristobal, Chiapas. Founded in the early 1950s by a former FBI agent named Thomas Stuart Ferguson, the foundation initially concentrated its work on the preclassic period, roughly 600 B.C. to A.D. 300, which, not coincidentally, corresponds to Book of Mormon times. Yet the foundation has hired many non-Mormon scholars over the years and has published its findings without a whiff of religious bias.

Likewise, Brigham Young University boasts a number of world-renowned Mesoamerican archaeologists, such as John Clark, who has done pioneering work in the area of the early-preclassic Maya, and Ray Matheny, whose National Geographic–funded excavations of the Mayan El Mirador ruins in the Peten rain forest of Guatemala are among the most extensive in the New World. Richard

Hansen, a Mormon archaeologist affiliated with UCLA, has digs under way elsewhere in the Peten that are already yielding intriguing finds.

Yet over the years southern Mexico has also seen a fairly steady procession of Mormon cranks and amateurs nursing zealous hopes of discovering the tomb of Nephi or the lost city of Zarahemla. Along the edges of legitimate Mormon-financed archaeology, one finds a colorful demimonde, one that has turned out a steady crop of grainy videos and specious books written in the sweeping style of Erich von Däniken's *Chariots of the Gods?* A number of resourceful travel operators from Utah have capitalized on the trend, leading Mormons on "Holy Land" package tours to the ruins of Mexico, running advertisements in the *Salt Lake Tribune* and the *Deseret News.* Hundreds of Mormons make these freelance trips each year, packing into sour-smelling buses, wielding machetes and metal detectors and occasionally an archaeologist's trowel. With neither academic credentials nor official permits allowing them to go digging for relics, they bushwhack through the rain forests and savannas of Central America on the scent of lost Semitic civilizations.

Inventing the Map

Mainstream archaeologists have scoffed at the church's long and, for the most part, discreet involvement with Mesoamerican archaeology — calling the Mormon theories patently absurd, procedurally flawed, even racist. The Smithsonian's National Museum of Natural History and the National Geographic Society have been so besieged with inquiries from enthusiastic Mormons over the years that both institutions have had to issue formal disclaimers stating that the Book of Mormon is not a historical text, and that no evidence points to the existence of a Jewish civilization in ancient America. Perhaps the most outspoken critic of Mormon archaeology has been Yale University's Michael D. Coe, one of the world's preeminent scholars of the Olmec and the Maya. The author of the best-selling book *Breaking the Maya Code,* Coe says there's not "a whit of evidence that the Nephites ever existed. The whole enterprise is complete rot, root and branch. It's so racist it hurts. It fits right into the nineteenth-century American idea that only a white man could have built cities and temples, that American Indians didn't have the brains or the wherewithal to create their own civilization."

Today, the ten-million-member Church of Jesus Christ of Latter-Day Saints is generally considered the fastest-growing denomination in the Western Hemisphere, especially among the Indian populations of South and Central America whose ancestors built the cities and temples that have so intrigued Mormon scholars. This is no accident, of course; the church has spent considerable money and effort proselytizing among the present-day Maya and other natives of the region, with church literature sometimes suggesting that the ancient Mexican god Quetzalcoatl was actually the triumphant Jesus Christ visiting the New World as depicted in the Book. Church missionaries often float the notion that American Indians are direct descendants of Book of Mormon peoples and are thus blessed with a sacred lineage.

Mormonism, in a sense, was born out of an inspired act of archaeology, Smith's stirring claim of having unearthed the golden plates. And to this day, the Book of Mormon remains a sacred text with a unique status, in the sense that its value and weight, its purchase on the imagination of the convert, crucially depend upon its acceptance as an authentic artifact of archaeology, a written work that is historically accurate and even testable. From its opening page, the Book of Mormon presents itself not as a sacred allegory but as the record of an extinct race of Hebrews who lived and sweated and died on real American soil. The events in the Book *had* to have happened, and somewhere on these shores, or the book is a fraud. Joseph Smith understood that any people with the sophistication of the Nephites surely would have left tangible traces of their civilization behind — a Hebrew inscription, a metal sword, a ruined temple mailed in jungle vine — and he always said that excavation work would eventually vindicate everything printed in the Book.

But over the past fifty years, as Mormon scholars have begun to apply the techniques of modern archaeology, the search has only grown more complex, more desperate, more discouraging. Adherents of other faiths and sects have, of course, encountered similar problems when the astringent of science has been applied to their most cherished beliefs. The fields of geology and paleontology, for example, do little to substantiate the truncated time line of the creationists — quite the contrary. Despite the painstaking efforts of numerous Christian archaeologists, not a shred of evidence has yet been produced that suggests the presence of Noah's ark on Mount Ararat in Turkey. For years, India and Nepal have been engaged in a rancorous and ultimately futile archaeological rivalry to resolve the ancient debate over which of the two countries was the true native land of Siddhartha (the Buddha).

Then again, the Book of Mormon does pose unique problems for the empirically minded reader — most fundamentally, the problem of a wholly hypothetical geography. Unlike a Holy Land archaeologist who can set up a dig in Jericho or Bethlehem and know with reasonable certainty that at least the location is about right, a Mormon archaeologist is forced to work from a map constructed entirely from guesswork: none of the Book's place-names match up with present-day sites, and the Americas lack the continuity of culture and language that one finds in Israel.

As archaeological digs throughout the Americas have increased our knowledge of ancient civilizations and led to such advances as the cracking of the Mayan hieroglyphic code, Mormondom has been forced to confront the problem of evidence. What happens when the ground refuses to cooperate, when the soil fails to yield what the faith insists is there? For many Mormons it's been a perilous quest, and more than a few who have ventured too far down the path have come back with their convictions in tatters, despairing at the lack of hard proof, wondering why the square pegs of belief won't fit into the round holes of the targeted terra firma.

The Golden Plates

Joseph Smith was a tall, rangy, young farmer when he began the arduous, two-year task of translating the Book of Mormon, "An Account Written by the Hand of Mormon upon Plates Taken from the Plates of Nephi." These golden plates, Smith said, were inscribed in an obscure hieroglyphic language called "Reformed Egyptian," which he was able to decipher only with the help of magical stones given to him by the angel Moroni. A long and densely written epic that Mark Twain later described as "chloroform in print," the Book of Mormon was published in 1830. Shortly thereafter, a new religious sect was born, the Church of Jesus Christ of

Latter-Day Saints. Smith and his followers moved west to Kirtland, Ohio, then west again, to the Illinois banks of the Mississippi River, where a little theocratic city called Nauvoo rose from the canebrake, with Smith as general, mayor, newspaper editor, social chairman, lodge wizard, and beloved prophet. He improvised his own little satellite world, his own frontier phratry, out on the edge of America. He took thirty wives. He commanded what was then the second-largest standing army in the United States. He steamed up and down the Mississippi in his private sternwheeler. He held grand feasts, dances, and wrestling matches. Smith was the life of his own party, following his passions right up until the end.

His most consuming passion, however, was for the American landscape itself — its ghosts and artifacts, the aboriginal prehistory of the New World, the puzzle of where the American Indians originated. In his youth, Smith had poked around the backwoods of New England as a "money digger," hunting for buried treasure that he said had been left by ancient civilizations. Throughout his life he was fascinated by Indian mounds and liked to spin intricate romances about who built them, and why. "Joseph would occasionally give us some of the most amusing recitals that could be imagined," the prophet's mother, Lucy Smith, once recalled. "He would describe the ancient inhabitants of this continent, their dress, mode of traveling, and the animals upon which they rode; their cities, their buildings, with every particular; their mode of warfare; and also their religious worship. This he would do with as much ease, seemingly, as if he had spent his whole life with them."

When news of the stunning Mayan ruins at Palenque reached the United States in 1841 with the publication of John Lloyd Stephens's *Incidents of Travel in Central America, Chiapas, and Yucatan,* Smith speculated that the Maya must have been Book of Mormon peoples. At one point he enthusiastically stated that the Palenque ruins were "among the mighty works of the Nephites." A Nauvoo newspaper article later attributed to Smith went on to suggest, "It will not be a bad plan to compare Mr. Stephens's ruined cities with those in the Book of Mormon."

By that time, however, Smith was already enmeshed in more pressing plots and subplots — his run for the U.S. presidency in 1842, controversies arising from the church's views on polygamy, and mounting squabbles with state and federal authorities. Then in 1844, at the age of thirty-nine, Smith was murdered by a lynching mob at a jailhouse in Carthage, Illinois, where he had been temporarily imprisoned on conspiracy charges. Several years later, the church began the exodus west under the stern gaze of Brigham Young, a stout man who proved to be a shrewd institution builder. Upon seeing the parched country around the Great Salt Lake, Young is said to have solemnly proclaimed, "This is the place!" To which his aide-de-camp responded, "Are you sure, Brother Brigham, are you sure?"

For the next hundred years, the church rarely revisited the question of just where in the New World the Nephites were supposed to have lived. The Book offered few clues. The place-names that cropped up in the text — Desolation, Manti, Shemlon, Bountiful — matched up with neither ancient Indian nor modern American geography, and the descriptions and coordinates were vague, at best. The Book spoke of a "Land Northward," which the church fathers generally guessed to be North America, a "Land Southward" (South America?), and a "Land of Many Waters" (the Great Lakes?). Given these parameters, the faithful were left to assume that the action in the Book had taken place in both North *and* South America, though mostly around upstate New York (especially the great Nephite-Lamanite battle depicted at the end), since that's where Smith had excavated the plates.

But within the anthropology department at Brigham Young University, another geographic paradigm began to evolve about fifty years ago. The more precisely scholars like BYU anthropologist M. Wells Jakeman studied the text, the more they realized that the action was, in fact, limited to an area of just a few hundred square miles. And the more they tried to superimpose the Book's mountains, rivers, oceans, weather, estimated travel times, and other characteristics over the physical landscapes of the Americas, the more apparent it became that wherever those few hundred square miles were, they certainly weren't anywhere near upstate New York.

When they boiled it down, what Mormon scholars were looking for was a "narrow neck of land," as the Book calls it, an isthmus set in a tropical climate (the text makes no mention of cold weather or snow) and surrounded by terrain known to have supported ancient peoples of sophisticated means — written language, masonry, astronomy, metal-working skills, and so forth. It eventually dawned on the scholars that they were throwing a dart at only one place, the same beguiling turf that Joseph Smith had speculated on from afar more than a century before: Mesoamerica, home of the Maya, the Olmec, the Toltec, the Zapotec, the Aztec, and other advanced civilizations of antiquity. (Not that Mormon scholars were arguing that the Nephites necessarily *were* the Maya or any of these other peoples; rather, that the Nephites had likely influenced them or were related to them in some way.) After much study, Mormon scholars narrowed their focus to an area that encompassed slices of Guatemala and Honduras, and parts of the Mexican states of Veracruz, Tabasco, Oaxaca, and Chiapas.

"Book of Mormon Lands," they called it.

Lost in Translation

If there is a "headquarters" for Mormondom's multifaceted interest in ancient Mesoamerica, it is a private nonprofit think tank called the Foundation for Ancient Research and Mormon Studies (FARMS). Housed on the BYU campus and handsomely endowed by the university and by faithful donors such as Mormon technobaron Alan Ashton (who founded the WordPerfect Corporation), FARMS is an energetic outfit that promotes all sorts of scholarship and research junkets. When I first called FARMS, for example, I was told that several FARMS researchers had proposed conducting "aerial reconnaissance missions" over southern Mexico to look for undiscovered ruin sites using the same "ground-pene-trating radar technology," developed at BYU, that the U.S. military used to peer into Saddam Hussein's bunkers. Here, pro-church scholars write spirited disquisitions on themes related to the antiq-uity of the Book of Mormon and publish apologetic books and pamphlets at an impressive clip. It's a kind of all-purpose clearing-house, the place inquisitive Mormons turn to for answers when critics raise nettlesome questions about the ancient provenance of the Book or the apparent paucity of archaeological evidence for Nephite civilization.

When I dropped by FARMS on a bitterly cold and gusty winter day, a middle-aged photographer who had just returned from a long trip across Mesoamerica was presenting a rather specious slide-show lecture to a small audience of faithful Mormons, a lecture that one of the more serious FARMS researchers would later describe as "the height of naiveté." "Look at that face!" the photographer said at one point, pausing the projector on a certain face from a Mayan relief at Tikal. "That's not an American Indian face. See the nose? That's not a nose characteristic of the area. That's a Semitic nose! And look closely. You see? He has a beard. What's that beard doing there? Well, that's interesting, because the Indians down there don't have facial hair. Where'd that beard come from?"

I was later led down the hall and introduced to the venerable white-haired theoretician sometimes referred to as the Thomas Aquinas of ancient Mormon studies — a tall, thin, precise gentleman in his mid-seventies named John L. Sorenson. A former chairman of BYU's anthropology department, Sorenson is a full-time scholar at FARMS and the author of numerous books, including the definitive work on the subject, *An Ancient American Setting for the Book of Mormon*. Personally involved in nearly every debate of consequence in the field for the past half century, Sorenson is one of the principal architects of the notion that the action of the Book of Mormon occurred in Mesoamerica. His first field trips to southern Mexico in the early 1950s set the tone and geographical parameters for much of the Mormon-affiliated research that has followed.

Ushered into his office, I found Sorenson leaning against a map of Mexico, absorbed in thought as he peered out his window at a winter storm sailing in fast from the alkali flats to the west. Once I sat down he snapped from his reverie, like a maestro satisfied that the crowd was now sufficiently hushed.

"You know," he began, "I've never asked the question, 'Did the events in the Book of Mormon happen?' I was born and raised in the church, and so for me this is beyond doubt. The question I've asked over fifty years of scholarship is, 'How did they happen?' Where did these people live, what were they like, what did they eat? I am very interested in establishing the Book's historicity. This is supposed to be the authentic record of a dead people. It won't suffice to say that Joseph Smith merely wrote it to impart a few spiritual truths. If it were ever conclusively demonstrated that Smith simply made it up, I don't know whether the church could survive."

Driven by this sense of spiritual urgency, and possessed of a polymath's grasp of interdisciplinary detail, Sorenson has spent the better part of his life hunkered in libraries, examining all sorts of arcane topics: linguistic cognates, ancient seeds of grain, comparisons of intestinal parasites, the possible resemblance of a specific Mayan glyph to a specific Hebrew character, and the insufficiency of the Bering Straits land-bridge theory to explain how *all* Native Americans arrived in the New World. Listening to Sorenson tick off these baroque lines of inquiry, I felt as though I were in the presence of a first-rate mind that had long since become inured to the stalemates and disappointments of a bedeviling scavenger hunt. "I've been at this for over a half century," he said, "and believe me, I have ways of managing the data reasonably so that I can take into account every apparent problem and contradiction in the Book."

The problems and contradictions in the Book are legion, in fact, and dealing with them has kept Sorenson and his colleagues ceaselessly busy for decades. Take the problem of elephants, to raise one prominent example. The Book mentions elephants several times, and yet as far as we know there weren't any elephants in Central America. This issue leads down a trail littered with imponderables: Could it be an error in translation? Could a woolly mammoth qualify as an elephant? Did mammoths ever exist in Central America, and at a time contemporaneous with Book of Mormon peoples? (So far, the evidence is no.) Should the church dispatch archaeologists to Mexico to hunt for mastodon bones?

The Book of Mormon describes dozens of other species of animals and domesticated plants that have yet to turn up in any pre-Columbian Mesoamerican excavations, including horses, asses, bulls, goats, oxen, sheep, barley, grapes, olives, figs, and wheat. This is not to mention all the inanimate objects: coins, functional wheels, metal swords, brass armor, chariots, carriages, glass, chains, golden plates.

The cumulative effect of all these minute examples would seem to deal a deathblow to the whole enterprise of Mormon archaeology. Yet BYU scholars like Sorenson have found all sorts of exotic rationales to circumvent these issues. Sorenson has gone so far as to postulate that the Book may actually have been referring to a tapir or a deer when Joseph Smith copied down the word "horse," although on the face of it, the idea of soldiers riding tapirs into battle seems ludicrously impractical. Sorenson has also suggested in his books and essays that the "chariots" referred to in the

Book weren't what *we* think of as chariots, but some considerably more primitive conveyance without wheels, more akin to a sled or a sledge, or even a nuptial bed.

Other Mormon scholars have been less willing to trowel over these apparent inconsistencies. In at least one public forum, BYU archaeologist Ray Matheny has been surprisingly blunt about the serious dilemmas posed by these rather glaring holes in the archaeological record. "I'd say this is a fairly king-sized problem," Matheny observed at a tape-recorded symposium in 1984 in Salt Lake City. "Mormons, in particular, have been grasping at straws for a very long time, trying to thread together all of these little esoteric finds that are out of context. If I were doing it cold, I would say in evaluating the Book of Mormon that it had no place in the New World whatsoever. It just doesn't seem to fit anything that I have been taught in my discipline in anthropology. It seems like these are anachronisms." Matheny concluded his talk with a sockdolager: "As an archaeologist," he said, "what [can] I say . . . that might be positive for the Book of Mormon? Well, really very little." Several Mormon archaeologists told me that Matheny's remarks caused considerable stir within church circles and came close to costing him his tenured position at BYU. Matheny has since carefully refrained from further public commentary on this subject, and he declined to be interviewed for this story.

Yet in 1993 Matheny's wife, Deanne G. Matheny, also a Mesoamerican anthropologist, echoed her husband's remarks in an essay entitled "Does the Shoe Fit? A Critique of the Limited Tehuantepec Geography." In taking Sorenson's elaborate apologetics to task, she wrote, "There are too many areas where one must either assume that evidence exists but has not yet been found or that something other than the words actually used [in the Book of Mormon] were intended. . . . Too much sidestepping of this sort can lead to the absurd."

With Sorenson's elastic style of argumentation setting the overall tone, there is about FARMS a dizzying buzz of intellectual energy, with scholars investigating every imaginable cranny of inquiry, from hermeneutics to meteorology, from animal husbandry to the prevailing currents of the oceans. Yale's Michael Coe likes to talk about what he calls "the fallacy of misplaced concreteness," the tendency among Mormon theorists like Sorenson to keep the discussion trained on all sorts of extraneous subtopics (like tapirs and nuptial beds) while avoiding what is most obvious: that Joseph Smith probably meant horse when he wrote down the word *horse,* and that all the archaeology in the world is not likely to change the fact that horses as we know them weren't around until the Spaniards arrived on American shores.

"They're always going after the nitty-gritty things," Coe told me. "Let's look at this specific hill. Let's look at that specific tree. It's exhausting to follow all these mind-numbing leads. It keeps the focus off the fact that it's all in the service of a completely phony history. Where are the languages? Where are the cities? Where are the artifacts? Look here, they'll say. Here's an elephant. Well, that's fine, but elephants were wiped out in the New World around 8000 B.C. by hunters. *There were no elephants!"*

Another eminent Mormon archaeologist of Mesoamerica, Gareth Lowe, has come down hard on Sorenson's attempts to, as he puts it, "explain the unexplainable." "A lot of Mormon 'science' is just talking the loudest and the longest," says Lowe. "That's what Sorenson is about, outtalking everyone else. He's an intelligent man, but he's applied his intelligence toward questionable ends."

Sorenson is quite well aware of his pariah status among non-Mormon archaeologists as well as in certain Mormon circles, and in a way he seems to relish the intellectual combat. He and his prolific, steadfast colleagues at FARMS are the last of the true believers, still confident that hard proof of Mormonism's essential truth will eventually emerge from the ground.

"This is a very, very lonely line of work," Sorenson conceded, running a hand through his thinning hair. "Non-Mormon archaeologists and anthropologists don't want to have anything to do with us. Still, Mesoamerica is such a wide-open field, with so many complexities and conundrums. Only one one-hundredth of one percent of the material has been excavated. And so I have complete faith that over time, the answers are going to rise up out of the forest carpet . . . like wild mushrooms."

Sorenson turned for a moment to watch the snowflakes that were tumbling outside his window. Suddenly, a vent opened in the clouds, and for a moment the Wasatch Mountains appeared, glowing pink as bubblegum over Provo.

The Sacred Mountain

Not long after my meeting with John Sorenson, I contacted a group of young Mormon financial consultants from Salt Lake City about to embark on their own two-week archaeological junket in southern Mexico. Merrill, Steve, and Jayson were close friends and business partners, all in their mid-thirties. With FARMS helping with some of their traveling expenses, they were heading down to survey a number of impressive ruins, from Monte Alban to Palenque to Chiapa de Corzo. I met them on a paintball field on the outskirts of Salt Lake City, where they held a weekly battle in Technicolor, and after the skirmish was over, they asked me to accompany them on their trip. I would be the fourth and only non-Mormon member of their "expedition," which was a bit of an overstatement, since they were without government permits and knew virtually nothing about the discipline of Mesoamerican archaeology. They called me "the gentile."

A few weeks later, we were renting a VW bus at the airport in Mexico City and heading for points south. We looped through the foggy, pine-forested highlands of Chiapas, still seething with its Zapatista rebellion. We met with dirty-nailed Mormon archaeologists in San Cristobal, nosed around in caves, and took a dory up the Rio Grijalva, thought to be the holy river "Sidon" that figures prominently in the Book of Mormon. The primary target of our trip, however, was the Olmec country along the Gulf Coast of Veracruz State. The rationale behind this focus had everything to do with John Sorenson. After much searching, Sorenson has postulated that a certain mountain along the coastal plains of Veracruz called Cerro El Vigia is the "most likely candidate" for the Hill Cumorah of the Book of Mormon. (As fantastic as it may seem, Sorenson actually argues that there were *two* Cumorahs: one in Mexico where the great battle took place, and where Moroni buried a longer, unexpurgated version of the golden Nephite records; and the one near Palmyra, New York, where Moroni eventually buried a condensed version of the plates after lugging them on an epic northeastward trek of several thousand miles.) Located between the little towns of Santiago Tuxtla, Santiago Andres, and Catemaco, Cerro El Vigia is the nub of a long-dormant volcano, hanging over pastures of Brahman cattle and sugarcane fields. My comrades' plan was to climb Cerro El Vigia — "the sacred mountain," they called it — with shovels and sifting crates and look around for evidence of the gory battle that may or may not have taken place there fourteen centuries ago.

Steve was an anxious, flaxen-haired chili-pepper fanatic whose mind constantly raced with pet conjectures fed by topo maps and dog-eared Mormon archaeology books. Jayson, on the other hand, was soft-spoken, skeptical, his deep brown eyes pooling with doubts about the advisability of the trip. "I can't help wondering where this all leads," he had confided in me, raising his voice to be heard over the howler monkeys thrashing in the surrounding canopy as we sat atop the Temple of the Inscriptions at Palenque. "I guess my logical requirements are more stringent."

It was Merrill who would turn out to be the natural leader of our expedition. Brash, fearless, a large guy with a knack for accelerating the plot of whatever situation in which he happened to find himself, Merrill had been dreaming about this trip for years, and his expectations were sky-high. "We're not here just to eat some tacos," Merrill had told me as he climbed aboard our rental bus in Mexico City. "We're all stalwart members. This is our Holy Land tour."

The night before our planned "assault" on the hallowed mountain, Cerro El Vigia, we stopped off in the nearby lakeside hamlet of Catemaco, a town famous all over Veracruz as an annual gathering place for witches and warlocks. We ordered a paella dinner at an outdoor restaurant and began to discuss the great Nephite-Lamanite battle on the Hill Cumorah. Merrill read to us from Mormon 6:7: "All my people had fallen; and their flesh, and bones, and blood lay upon the face of the earth, being left by the hands of those who slew them to molder upon the land, and to crumble and to return to their mother earth."

"It was a bloodbath," said Steve. "Hundreds of thousands of Nephite corpses. Any battle that big, there's bound to be local legends."

"Exactly," said Merrill. "So what we need to do is find the head *brujo* of Catemaco and plumb his knowledge of the folklore around here. They say his name is Apolinar. Supposed to be the most famous one in Veracruz State. He lives just down the street here, at Hidalgo number twenty."

We eventually found the house, just off the *zócalo*, and studied the little sign out front — "Apolinar Gueixpal Seba, Botánica y Ciencias Ocultas." Merrill rapped on the massive oak gate. After a long wait, the hinges creaked open, and there stood Apolinar himself. He was a frightening sight, a Hispanic version of Alice Cooper, attired in black leather pants and a black leather vest draped over luridly tattooed pectorals. He seemed unhappy to see us, as if we'd just interrupted something — the weekly infanticide, perhaps. But when Merrill paid him something in advance for his services, Apolinar reluctantly led us back to his lair, a slatternly room crowded with jarred elixirs and dried insects and the mingled fragrances of a dozen incense sticks.

"So, may I help you in finding a loved one?" Apolinar's eyes glinted in the thin light of a votive candle. "Or are you ill?"

"No, *gracias*," said Merrill, who speaks fluent Spanish from his days as a missionary in Guatemala in the late 1980s. "We are Mormons. We've come from Utah, in the United States, to learn about Cerro El Vigia."

Apolinar regarded us in silence for a long moment, and said, "Ah, El Vigia. It is a magical place."

"Magical in what way?" Merrill asked.

"There are so many legends. It is said that there was once a fierce and bloody battle."

"A *what?*" Merrill said, his interest quickening.

"*Sí*, it is an old, old story," Apolinar went on. "Hundreds and thousands fell. It is believed that their ghosts are still up there, swirling in the mist."

Merrill was hungry for more. "In this battle you speak of — who was doing the fighting?"

"I cannot say more. It is a belief we do not like to discuss. But, if you must know more, well . . . it is said that there is a book buried up on the mountain."

Merrill was beside himself. A book? Buried in the hill? This is precisely what the Mormons believe — and I could see that Apolinar knew it. He'd doubtless heard the story of the Mormon interest in Cerro El Vigia before, and he'd seen those squads of clean-cut missionaries all suited up and knocking on doors around town. I sensed that he was perhaps preying on Merrill's hopes a little, just for kicks.

Apolinar could have been Lucifer himself, but Merrill seemed buoyed by everything the *brujo* had said. After we left Apolinar's place, Merrill drove us toward Cerro El Vigia, which was faintly visible in the moonlight, a dim swell of basalt scarved in fog. Merrill said he'd made up his mind to buy a little hacienda in the town of Santiago Tuxtla so he could come down from Utah on a regular basis to live near the sacred mountain. Now, as he beheld its presence, there was a look of misty awe in his eyes, the same devout look I'd seen on the faces of the Mormon pilgrims up in Palmyra, New York. It was the sentimental gaze of ancestral longing, the yearning for a kind of motherland. Only this was a motherland based on literary constructs and anthropological speculation rather than on bloodlines, a theoretical motherland thrice removed, with Hebrew ancestors said to be related to American descendants through an Egyptian-language text purportedly unearthed over 150 years ago by a young farmer nearly three thousand miles north of here. It was a nostalgia, in other words, that had to travel through a fabulous, labyrinthine circuit before it could be felt.

Letting the Ground Speak

It's doubtful that any Latter-Day Saint has ever felt this sense of sentimental kinship with the Nephites as vividly as the late Thomas Stuart Ferguson, an attorney and former FBI agent who, from the early 1950s to the 1970s, was more or less the godfather of Mormon archaeology. Born in Pocatello, Idaho, and educated at Berkeley, Ferguson was a vigorous, headstrong man who believed with absolute certainty that excavations in Mexico would one day vindicate the Mormon faith. In the late 1940s, flush with excitement over the new Mesoamerican parameters that had been staked out by BYU scholars, Ferguson personally tromped through the jungles of Chiapas hunting for suitable candidates for Nephite ruins.

One of his comrades on those early freelance expeditions to Mexico and Guatemala was his friend J. Willard Marriot, the hotel magnate. In one letter, Marriot recalls, "We spent several months together in Mexico looking at the ruins and studying the Book of Mormon archaeology. I have never known anyone who was more devoted to that kind of research than was Tom."

Another of Ferguson's traveling companions to Mexico was John Sorenson, who was then a young anthropology Ph.D. candidate. "Tom was a lawyer, first, last, and always," Sorenson told me. "He had no training in archaeology. To him, things had to be *proven*. He wanted to hit the jackpot, to find a chariot or a Hebrew inscription or something. He was betting everything on a pull at the slot machine. Ferguson's view was, the Book of Mormon talks about horses, there should be figurines showing horses. So everywhere he went, his first question to campesinos was 'Seen any figurines of horses?' Tom felt like he had to have something moderately spectacular to sell to the church. No archaeologist had ever systematically looked at Chiapas before, so we took a Jeep up there and looked around." The results were impressive: Sorenson and Ferguson were able to identify some seventy potential sites in less than two weeks of traveling.

In 1952 Ferguson formed the New World Archaeological Foundation and then set about soliciting funding from the church and from well-to-do Mormon benefactors. "If the anticipated evidences confirming the Book of Mormon are found," he wrote in a letter to David O. McKay, the president of the church, "world-wide notice will be given to the restored gospel through the Book of Mormon. The artifacts will speak eloquently from the dust."

In another letter to McKay, Ferguson wrote, "The source of our income and support for the work can be kept strictly confidential if it is desired . . . [but] the Church cannot afford to let all of the priceless artifacts of Book of Mormon people fall into other people's hands. We can make wonderful use of them in missionary work and in letting all the world know of the Book of Mormon."

Finally, in 1953, President McKay relented, and the church quietly presented the New World Archaeological Foundation with an initial grant of $15,000, with a much larger sum of $200,000 to be given in 1955. Ferguson was shrewd enough to realize that if his quest were to succeed, he must hire objective, non-Mormon scholars, and he lured some of the most prominent names in the field, including Gordon F. Ekholm, who later became curator of American archaeology at New York's Museum of Natural History, and A. V. Kidder, the grand old man of American archaeology. From the outset, Ferguson stipulated that the NWAF "would not discuss direct connections with the Book of Mormon, but rather [would] allow the work to stand exclusively on its scholarly merits."

"Let the evidence from the ground speak for itself," Ferguson declared, "and let the chips fall where they may."

The NWAF set up its first large dig at Chiapa de Corzo, and the site proved a fabulous trove for studying the formative preclassic period. Ferguson was ecstatic. "The importance of the work carried out this past season cannot be overestimated," he wrote in a letter to the First Presidency of the church. "I know, and I know it without doubt and without wavering, that we are standing at the doorway of a great Book of Mormon era." In 1958, in an enthusiastic and notably amateurish survey of Mesoamerican archaeology titled *One Fold and One Shepherd*, Ferguson wrote, "The important thing now is to continue the digging at an accelerated pace in order to find more inscriptions dating to Book-of-Mormon times. Eventually we should find decipherable inscriptions . . . referring to some unique person, place or event in the Book of Mormon."

In October of 1957, NWAF archaeologists dug up a cylinder seal from a site at Chiapa de Corzo that caused immediate excitement. The seal was inscribed with an unusual-looking ornamental design that, to Ferguson's eyes at least, resembled Egyptian hieroglyphics. In May of the following year, he sent a photograph of the seal to an eminent Egyptologist at Johns Hopkins University named Dr. William F. Albright. Without prompting from Ferguson, Albright examined the photograph and, in a letter, stated that the cylinder seal contained "several clearly recognizable Egyptian hieroglyphs." Although other Egyptologists would later dispute Albright's assessment, Ferguson was overjoyed, believing with heart and soul that this was the first piece of incontrovertible proof of the Nephites. "In my personal opinion," he wrote in a moment of religious abandon, "[Albright's finding] will ultimately prove to have been one of the most important announcements ever made."

An Artifact Discovered

The rutted dirt road on the back side of Cerro El Vigia winds through green-black jungle, past the tin-sided shacks of campesinos, and eventually peters out on the high, wind-scrubbed flanks, where thousands upon thousands of enormous basalt boulders are spread over the golden grass like caviar on toast. These are the lava fields that provided the raw material for the colossal Olmec busts — some of which weigh more than ten tons — that now squat in town squares along the Veracruz coast. How they managed to drag these immense rocks from the mountains is one of the many riddles that surround the Olmecs, who died out around 400 B.C. and are generally considered the progenitors of all other advanced civilizations in Mexico.

Steve, Jayson, and I were standing amid this boulder field, while Merrill held a compass in his hand and surveyed the landscape like a commanding general, envisioning the battle lines as they must have looked during the great Nephite-Lamanite engagement. We had been up here all day, wandering through a maze of impressive petroglyphs. It was dusk now, and Mexican free-tailed bats swooped down at us, attracted to the bugs that were attracted to our headlamps. Down in the valley, the first lights of Santiago Tuxtla gave off a skim-milk blue.

In the gathering darkness, a campesino named Carlisto pointed out a long, slender boulder lying in the scrub. On its underside, he said, there was rumored to be an elaborate carving that dated back to Olmec times. Apparently it had fallen over years ago like a pillar at Stonehenge, and no one had ever bothered to right it.

Merrill stood there considering the capsized monolith. He brushed his hand over the hard, pebbly surface and scanned it with his flashlight.

Maybe, Carlisto politely suggested, we would like to come back tomorrow morning and have a better look?

"I say we turn it over right now!" Merrill replied, and as if to emphasize his point, he shined his flashlight in our faces. "We've got plenty of manpower here," he added, nodding at the dozen or so friends and relatives of Carlisto, who'd gathered to see what the commotion was about.

Presently, all of us assumed our places around the rock and started building up a rhythm of shoves, tossing in stone chocks after each heave, while Merrill used a large log as a prying lever. Soon we could see a piece of the underside, but it was caked in dirt and hard to make out.

"Maybe it's a horse," Steve said, hopefully.

With one last push, the boulder tipped forward and tumbled downhill. Twenty yards below us, it rolled to a stop in a cloud of dust. We all scurried over to it. There was just enough juice left in Merrill's flashlight to limn the outlines: A round lobe here. Another lobe over there. A long shaft that culminated in . . .

The campesinos couldn't contain their laughter. It was impossible to ignore the obvious. After an exercise that only hinted at the hernias and slipped discs the Olmecs must have suffered as they hauled their titanic rocks to the coast, we had succeeded in unearthing what must be one of the most magnificent stone phalluses in the New World.

"What does this mean for the Book of Mormon?" asked Steve.

"It doesn't mean jack!" Merrill replied, laughing for a while with the others. Then, as the campesinos all wandered back to their shacks for the night, Merrill lingered in silence by the monolith, catching his breath, wondering whether this was, in fact, the place.

A World Upside Down

Despite Tom Ferguson's nearly effervescent zeal, the New World Archaeological Foundation somehow managed to hold fast to its original pledge to keep Mormonism out of its scholarship, and over the years it developed an international reputation for first-class work. This had much to do with the efforts of Gareth Lowe, the meticulous Mormon archaeologist who served as the foundation's director for thirty years. "We were always dealing with a tension between doing good scholarship and just digging for Mormonism," recalls Lowe, who is now retired and living in Tucson, Arizona. "The church would tell people in the congregation, relax, we have people down there who're investigating things. Just hold tight. They're on the case. But when I went down there, I realized I was very green and wide-eyed. I decided early on that we might never find anything that proves the Book of Mormon. But by doing good science, at least we could make a contribution. There was almost nothing known about these early cultures."

I asked Lowe whether, after all those years of digging under the auspices of the church, he was still a faithful Mormon. He paused thoughtfully for a long moment and then replied, somewhat gingerly, "Well, my wife still is."

Yale's Michael Coe worked with Gareth Lowe and other NWAF scholars in the fifties, sixties, and seventies, and says he has "nothing but absolute admiration" for their work. "They did the first really long-term, large-scale work on the preclassic in Mesoamerica, and they published it all. And by and large, their Mormonism never came through. Occasionally they'd get these dopes out of

Utah who'd arrive with metal detectors and earphones and march
around their sites trying to find the plates of gold. But the foundation's scholars always made sure they got on the plane and went
back home. What's amazing is that they were able to do this kind
of scholarship within the context of what is essentially a totalitar-
ian organization. There isn't much of a difference between the
old Red Square and Temple Square. But as in the Soviet Union,
even given the terrible theoretical framework that they had to
operate under, the foundation managed to do excellent work in
spite of it."

By the early 1970s, surveying all of the foundation's notable
findings, Thomas Ferguson began to assemble the case for the
Book's ancient origins. Other than the "Egyptian" cylinder seal, the
NWAF excavators had found nothing that seemed to authenticate
the Mormon faith. Ferguson grew increasingly alarmed by this lack
of progress. In a letter dated June 5, 1972, he would write, "I sin-
cerely anticipated that Book of Mormon cities would be positively
identified within ten years — and time has proved me wrong."

What began merely as a mild suspicion would become an inexo-
rable undertow of doubt. In 1975 Ferguson wrote a twenty-nine-
page paper analyzing the case for Mormon archaeology. Entitled
"Written Symposium on Book-of-Mormon Geography," it had all
the hallmarks of a legal brief. "With all of [our] great efforts, it cannot be established factually that anyone, from Joseph Smith to the
present day, has put his finger on a single point of terrain that was
a Book-of-Mormon geographical place. And the hemisphere has
been pretty well checked out by competent people. Thousands of
sites have been excavated." In a detailed chart that poignantly
illustrated his spiritual despair, he went on to enumerate all the
plants, animals, and artifacts mentioned in the Book of Mormon
that were as yet undiscovered in ancient Mesoamerican digs. Under the heading "Evidence supporting the existence of these forms
of animal life in the regions proposed," he ticked off: "Ass: None.
Bull: None. Calf: None. Cattle: None. Cow: None. Goat: None.
Horse: None. Ox: None. Sheep: None. Sow: None. Elephant: None
(contemporary with Book of Mormon). Evidence of the foregoing
animals has not appeared in any form — ceramic representations,
bones or skeletal remains, mural art, sculptured art or any other
form. . . . [T]he zero score presents a problem that will not go away

with the ignoring of it. Non-LDS scholars of first magnitude, some of whom want to be our friends, think we have real trouble here."

In this same legalistic fashion, Ferguson surveyed the long list of plants and artifacts that pose similar problems for the Book of Mormon: barley, figs, grapes, wheat, bellows, brass, breastplates, chains, copper, gold, iron, mining ore, plowshares, silver, metal swords, metal hilts, engraving, steel, carriages, carts, chariots, glass. The evidence for their existence in pre-Columbian Mesoamerica, he succinctly summarized, was "zero."

Eventually Ferguson, the indefatigable apostle and founder of Mormon archaeology, came to the anguished conclusion that Joseph Smith had simply invented the Book of Mormon out of whole cloth. He pronounced Mormonism a "myth fraternity," and slipped into a profound spiritual crisis that lasted until his death, of a heart attack, in 1983. "You can't set Book of Mormon geography down anywhere," he wrote in 1976, "because it is fictional and will never meet the requirements of the dirt-archaeology. What is in the ground will never conform to what is in The Book." And in another letter: "I have been spoofed by Joseph Smith."

Precisely when Ferguson lost his faith is not entirely clear — it seems to have been a gradual process, and he was very discreet — but his disillusionment dates at least as far back as December of 1970, when he paid a curious visit to ex-Mormons Jerald and Sandra Tanner, owners of Lighthouse Bookstore and Salt Lake City's best-known critics of Mormonism. "He sat in our shop and told us that he had lost faith in the historicity of the Book of Mormon," Sandra Tanner told me when I stopped by her bookstore. "This was just astounding to us. Tom Ferguson was the big answer man of Mormonism. He was the man who had gotten the church's hopes up. He'd said to the church, 'If the Book of Mormon really is history, we ought to be able to find something if we throw enough money and expertise at it.' He seemed grieved by the fact that he had wasted all those years of his life trying to prove the Book of Mormon."

Ferguson did not broadcast his disenchantment with the Book of Mormon, in large part because he had close family members who were still faithful and because he still enjoyed some of the church's social aspects. Consequently, his crisis of faith was not widely known within church circles. In 1990, however, the liberal Mormon journal *Dialogue* published a controversial essay titled "The Odyssey of Thomas Stuart Ferguson." Written by a University of Utah librarian named Stan Larson, the essay told the Ferguson story in its entirety for the first time. (Larson's essay has since been expanded into a book, *Quest for the Gold Plates.*) The long, and ultimately painful, arc of Ferguson's relationship with Mormon archaeology has had powerful resonance for a new generation of Mormon liberals, who have tried to reconcile what they view as major problems in the Book of Mormon with the latest findings of science and ancient scholarship.

This new line of revisionist thinking came to something of a crescendo with the publication, in 1993, of *New Approaches to the Book of Mormon: Explorations in Critical Methodology,* a much-talked-about collection of essays written mostly by apostate former Mormons and edited by a young Mormon-raised, self-taught scholar named Brent Lee Metcalfe. The book, which one reviewer went so far as to call "the most sophisticated critique of Mormonism to date," has been banned from all church-affiliated bookstores, and several of the book's contributors, including Metcalfe, have been formally excommunicated.

"There is a new wave of younger, savvier intellectuals who've come along in the wake of Ferguson's disillusionment who simply cannot square the Book of Mormon with the scholarship," Metcalfe told me when I met him at his home in a southern suburb of Salt Lake City. "In order to accept the Book of Mormon as a factual record, one has to be willing, literally, to turn one's whole world upside down. North is no longer north, south is no longer south, a horse is no longer a horse, and chariots don't have wheels. No other historical text would make these kinds of demands on its readers. If one has to go to all these tremendous lengths to make this book work, then what's the point?"

It Happened Someplace

Despite this new current of doubt within liberal Mormon intellectual circles, and despite its own patriarch's profound disenchantment, the New World Archaeological Foundation lives on today. It's a small, dedicated outfit based in San Cristobal, Chiapas, with a tiny staff of archaeologists still quietly digging in the dirt of southern Mexico. When I stopped by to visit the foundation, I was greeted by archaeologist Ron Lowe, Gareth Lowe's son, who gave me a tour of the musty offices and examining rooms, with topo maps on the walls and countless portfolio drawers filled with carefully cataloged potsherds and artifacts. The foundation's budget has been scaled back, perhaps because the church leaders saw in Ferguson's story a cautionary tale about the perils of using science to "prove" the historical origins of the faith, and perhaps because so little had been found to pique the faithful's interest. The scale-back came in the mid-1990s, shortly after the foundation staff was embroiled in an embarrassing sex scandal: one of the senior Mormon archaeologists was accused of sleeping with the under-aged daughter of an NWAF employee, and this allegation led to a number of firings and a wholesale rethinking of the foundation's mission.

Still, Brigham Young University remains committed to funding the NWAF, and its current director, the respected Mesoamericanist and BYU professor John Clark, has pursued a cautious course of serious, no-nonsense archaeology.

"Everybody still believes we've got this secret agenda to validate the Book of Mormon, and it makes my life very difficult," Clark told me. "The problem is, we have these so-called Book of Mormon tours, we have a lot of people running around trying to find Nephi's tomb. I get very nervous about people knowing more than they can possibly know. Archaeological data in the hands of the wrong person scares the heck out of me."

Clark spoke with all the concentrated caution of a high-wire artist. I could sense that he'd had much practice negotiating the fine line that's strung between the faith that sustains him, the university that pays him, and the scholarly discipline that gives him professional respect. He said he wished Mormon archaeology, as a subject, would go away. Yet it was more than mere coincidence that of all the regions of the world, he'd chosen ancient Mesoamerica as the place to sink in his trowel and stake his career for Brigham Young University. It was as though the ghost of Joseph Smith were perched on his shoulder, pointing enthusiastically at maps and continents, suggesting places to dig for the ultimate treasure. Clark did his best to tune him out, but the founder's ghost was such a steady distraction, proposing such quixotic goose chases, spinning such fanciful diversions, that it was virtually impossible to ignore his presence, try as one might.

"Look," Clark finally said, "I'm just trying to be a professional archaeologist. To me, the Book of Mormon has the feel of an ancient document, and any problems are problems of translation. I believe it did happen someplace. I just don't know where. But I, for one, can live with the uncertainty."

CRAIG B. STANFORD

Gorilla Warfare

FROM *The Sciences*

HIGH AMONG the Virunga volcanoes, along the eastern edge of
the Democratic Republic of Congo (DRC), there lives a group of
gorillas with little interest in international politics. Day by day and
week by week they wander through meadows of bracken fern, eating bamboo and nettles, mating in polygynous groups, and fastidi-
ously grooming one another. Although there are only around six
hundred mountain gorillas left in the world — half of them here
and half in Uganda's aptly named Bwindi-Impenetrable National
Park — the gorillas themselves seem unconcerned about that fact.
Their most aggressive, most territorial act toward people is to bite a
farmer on the behind now and again.

But if the forest, to a gorilla's eye, seems peaceful and unbroken,
from a human perspective it is riven by disputes, crosshatched by
historical, political, and biological borders. The volcanoes themselves may be dormant, but they straddle three of the most incendiary places on earth: the southwestern tip of Uganda, the eastern-
most edge of Congo (formerly Zaire and officially known as the
DRC), and the northernmost lip of Rwanda. The first has a history
of violent dictatorship; the other two are still shuddering with brutal conflicts. On any given day, therefore, the visitor cannot be sure
whether he will run into gorillas or guerrillas on the trail.

The gorillas don't care: every year they walk across the spine of
the volcanoes, spend several months in Uganda, then walk back
home to Congo. But the camera-laden Westerners who have hiked
for hours to watch them have no such luxury. Political instability
put an end to gorilla-viewing in Congo as well as in Rwanda. Tour-

ists are still free to watch the Bwindi gorillas, but when the Virunga gorillas reach the Congo border, the tourists have to stop and watch them disappear into the trees. Then the tourists turn and head back to camp.

This past March 1, however, that peaceful pattern was suddenly threatened. That morning a group of guerrillas plunged across the Congo border and into Uganda, in violation of international law, in search of unsuspecting ecotourists. Known as the Interahamwe, the guerrillas were Hutus from Rwanda who for years had fought the Tutsi minority that ruled their country. When war erupted in Rwanda in 1994, after a plane carrying the Rwandan president was shot down, the Interahamwe began a brutal campaign of genocide against the Tutsi. In the ensuing slaughter, carried out by both sides, often with machetes, between 10,000 and 20,000 Interahamwe were chased into eastern Congo, where dense forest and political chaos afforded them the best possible cover. There they regrouped and licked their wounds, stealing radios and other supplies to maintain their ragtag revolutionary army.

With Ugandan soldiers patrolling nearby, the Ugandan army did not suspect that the rebels would risk an attack, especially on Western tourists. But the Interahamwe were furious not only with the Rwandan government, they were also incensed with the Ugandan and Western governments, which had supported Rwanda. And so, on that March dawn, more than a hundred rebels charged into Bwindi-Impenetrable Park, armed with assault rifles, and headed toward the ecotourism center at Buhoma. Soon a warden was dead and fourteen Western tourists had been seized and taken hostage in the forest. Among them was Mitchell A. Keiver, a Canadian research assistant who worked with me at a nearby camp where I study gorillas and chimpanzees.

Until recently, mountain gorillas had good reason not to fear the region's bloody politics: they were worth far too much alive. Before neighboring Rwanda slid into genocidal chaos, gorilla tourism among the Virunga volcanoes was that nation's second-largest source of revenue, after coffee. As many as 1 million Rwandans were killed during the massacres, and 2 million more were forced to flee, but only five gorillas were killed by the military. No matter who ended up controlling Rwanda, both sides must have reasoned, the gorillas had to remain safe and sound.

Mountain gorillas owe much of their fame, and thus their value as tourist attractions, to the pioneering field research of the primatologist Dian Fossey and the dramatization of her life in the 1988 film *Gorillas in the Mist.* Fossey's work reversed the popular image of gorillas, born of countless Tarzan and King Kong movies. Rather than bloodthirsty, chest-thumping behemoths — testosterone in black leather — gorillas were depicted as the easygoing animals they are, threatened by people rather than threatening to them. They are herbivores, after all, and exceedingly shy ones at that. A four-hundred-pound male, unaccustomed to tourists, will bolt into the forest, trailing a stream of diarrhea, at the mere sight of a person.

Fossey's work, and the movie about her, sparked an international effort to protect the gorillas. But that effort, in some ways, depends on borders and classifications as arbitrary and disputed as the invisible line that separates Uganda from Congo. Like most animal and plant species, gorillas have traditionally been broken into subspecies, or races, by taxonomists. According to the current classification, there are three kinds of gorilla in Africa (though that number may soon rise), each with a different look, lifestyle, and habitat, and a dramatically different population status.

The mountain gorillas of Bwindi and the Virungas, officially known as *Gorilla gorilla beringei,* are a relict population that long ago migrated from the west and evolved in isolation. Much farther west are the western lowland gorillas, which make up a second subspecies, and compared with their montane cousins, there are lots of them. The western lowland gorilla was the first to be named and therefore carries the doubly redundant moniker *Gorilla gorilla gorilla.* Estimates of its population vary greatly, but a conservative count would put it between 100,000 and 150,000.

The rest of African gorilladom is represented by *Gorilla gorilla graueri,* the eastern lowland gorilla, with traits somewhere between those of the lowlanders to the west and the highlanders to the east. According to a recent survey carried out by the Wildlife Conservation Society, about 7,000 eastern lowland gorillas are left, all in war-torn Congo.

Lowland gorillas are smaller, grayer, and shorter-haired than gorillas in the mountains. Unlike their more sedentary cousins, lowland gorillas may travel miles each day in search of fruit. The lowland so-

cial system, in turn, seems to reflect that mobility: it would be unusual indeed to find mountain gorillas wandering far from the core
of their group, whereas lowlanders disperse on a daily basis.

All three gorilla subspecies are in trouble, but *G. g. beringei* is
clearly in the direst straits. In the past several hundred years, hunters, farmers, and loggers have painted the gorillas into a mountain-
top corner. The villages and farms that surround them are among
the most densely populated on the continent, leaving few resources for the gorillas; logging has reduced their habitat to a few
small patches of forest, and poachers have sold the babies to zoos
and chopped up the adults to sell as tourist trinkets. The six hundred that remain would be a paltry number for any species, but for
gorillas, which reproduce once every four years, it could be disas-
trously small.

Logging and hunting are closely linked in western and central
Africa. Every year nearly 16,000 square miles of forest are felled on
the continent. And as logging roads reach into the great forest
tracts, they give hunters with rifles access to apes that once had only
tribes with spears to fear. The logging companies themselves encourage hunting, which feeds their workers in forest camps, and
gorillas and chimpanzees are sold as table items to wealthy and
middle-class Africans, who see bush meat as a status symbol. "People don't want beef or chicken anymore — they call it 'white man's
meat,'" one conservationist told the *New York Times*. As a result, the
bush-meat trade kills some 3,000 lowland gorillas every year. Even
western lowland gorillas, with their population of 100,000, will not
withstand such pressures for long.

The one point of light in this bleak vista is ecotourism. In recent
years Uganda has recovered from its years of political chaos to become something of a model African country for wildlife conservation. The Ugandan Wildlife Authority has built a thriving industry
around the viewing of mountain gorillas. In Bwindi the country
boasts a UNESCO World Heritage site, as well as one of the most
beautiful forest landscapes on the African continent. In Buhoma,
the site of the now-infamous kidnappings, several camps were set
up for gorilla-watching. The fanciest of them, run by the world-renowned safari organization Abercrombie and Kent, took visitors
back in time to the Africa of Ernest Hemingway. Great green tents
sheltered tourists from the frequent rain, local cooks prepared ex-

otic feasts, and waiters in white shirts and bow ties served mixed drinks on hot afternoons. The only concession to modernity was that no animal was killed there. Tourists, as they say, shot only with their cameras, not with guns.

It is a scene to provoke a sneer, perhaps, from some conservationists. But the fact is, in Bwindi, ecotourism *works*. Poaching is under control and the remaining gorilla habitat is protected in national parks and reserves. So comfortable have the gorillas grown around people in recent years that they have taken to foraging in farmers' fields (hence the occasional butt bite), and they allow people to sit by them. That approachability, combined with the animals' rarity, their fame, and their relatively sedentary ways (compared to the lifestyle of their wide-ranging lowland cousins), have made the mountain gorillas into ideal tourist attractions.

Indeed, gorilla-watching has become the centerpiece of the modern tourist's African safari. People pay as much as $3,130 for package tours in which they slog through rugged, wet terrain in order to spend an hour a day sitting with a group of habituated gorillas. Multiply that by several groups of gorillas, each visited by six high-rolling tourists a day, and the tourist income adds up quickly. Every year the farmers who live around Bwindi and the Virungas, scratching out a living from the steep volcanic slopes, get 20 percent of the revenue collected from gorilla-tracking permits. And gorilla tourism heavily subsidizes the Ugandan Wildlife Authority's budget as well. The fate of the mountain gorillas, in other words, is inseparable from the fate of the people who live around them.

Thanks to ecotourism, mountain gorillas are among the best-protected primates in the world. But a double-edged sword hangs over the area, one edge political, the other taxonomic. Deprive the gorillas of their safeguarded wildlife preserves, and loggers and poachers will soon drive them to extinction. Change their biological classification, and their conservation status could be put in jeopardy. Both lines of protection, unfortunately, are in danger of being cut.

In 1996 the primatologists Esteban E. Sarmiento of the American Museum of Natural History in New York City and Thomas M. Butynski and Jan Kalina of Zoo Atlanta proposed a new classification scheme for mountain gorillas. The Bwindi gorillas, they argued, should be recognized as a new and separate subspecies from the Virunga gorillas. The suggestion made a certain sense. The Bwindi and Virunga gorillas already live in distinct groups, about twenty-five miles apart. Five hundred years ago, before subsistence farmers finished clearing the land between the two mountain ranges, the two gorilla habitats were one. Since the separation, however, the populations no longer mingle or interbreed, and their differing habitats — or, perhaps, the random accumulation of genetic mutations in both groups — have, over the centuries, wrought differences in the two groups' looks and behavior.

Because my research project began just after Sarmiento, Butynski, and Kalina published their paper, I was in a perfect position to observe those differences. From our camp — a collection of mud-and-thatch huts clinging to the steep side of a mountain — gorillas and chimpanzees can be heard in the forest below around the clock. And we spend most of our waking hours collecting information on the animals' habits. The Bwindi gorillas, we have found, tend to eat more fruit than the Virunga gorillas, and so they travel farther each day. Whereas the Virunga gorillas almost always sleep on the ground, the Bwindi gorillas sometimes sleep in trees. Most obviously, the Virunga gorillas have shaggier coats and a slightly larger build.

Are such differences enough to redefine the two populations as separate subspecies? That depends on your definition of the term. In 1942 the biologist Ernst Mayr defined a subspecies as a population of animals that has adapted genetically to a unique environmental niche and that reproduces in isolation from others of the species. Two populations of squirrels that live on opposite sides of the Grand Canyon might fit that definition, and so would the Bwindi and Virunga gorillas. But over the years the concept of subspecies has undergone an evolution of its own. Some biologists now define it genetically; others focus on social structure or the mechanisms of mate recognition; still others proceed mathematically, by grouping physical traits through cladistic analysis. In general, the various factions can be grouped into two larger camps — the "lumpers" and the "splitters" — but even members of the same camp agree on few specifics. Mayr himself, in 1982, recanted his earlier view, describing subspecies as merely convenient pigeonholing devices for taxonomists.

To most laypeople, genetic differences seem the most plausible way to divide up species. The longer two populations live and breed apart, the more distinctive mutations accumulate in their genes. What could be clearer and simpler? Yet speciation and evolutionary distance are not intimately connected. Two animal populations can look quite different from each other — Great Danes and Chihuahuas, for instance — yet remain closely related genetically. Or they can look very much alike — the orangutans from Borneo and those from Sumatra are a good example — even though they diverged from the same ancestral stocks many thousands or even millions of years ago.

Faced with so many variables, and so little agreement about how to weigh them, biologists have little choice but to use a scorecard approach. And that substantially complicates the classification question for mountain gorillas. In 1996 Karen J. Garner and Oliver A. Ryder, two geneticists at the Zoological Society of San Diego in California, compared the mitochondrial DNA of all the gorilla populations. Whereas mountain gorilla DNA is quite different from lowland gorilla DNA, they found, the DNA of the Bwindi and Virunga populations is virtually indistinguishable. How, then, should those groups be classified? Should they be put in separate subspecies, on the basis of their looks, behavior, and biogeography, or should they be kept together, on the basis of their genes?

In this case, I believe one must side with the lumpers. Genetic evidence may be no trump card in matters of speciation, but among mountain gorillas the physical and behavioral differences are even less definitive. Gorillas in the Virungas live more than 10,000 feet above sea level, in cold, misty meadows rich in herbs and ground cover but poor in fruit. Bwindi gorillas live in lower and warmer mountains, and so they have access to more fruit trees. Hence the physical circumstances alone can explain why Bwindi gorillas eat more fruit and travel farther to find it, whereas Virunga gorillas make do with foliage. Gorillas are smart, flexible primates, and they have simply adapted their behavior to local conditions.

The physical differences between the groups are just as easy to explain. When my cat and I moved from Michigan to southern California some years ago, she stopped putting on a layer of fat and thick fur every winter. That is because cats, like many animals, have an inherent ability to respond to low nighttime temperatures, decreasing daylight hours, and other signs of approaching cold. Gorillas can no doubt respond to their environment in a similar way.

As abstract and academic as the speciation debate seems, it could become a matter of life or death for mountain gorillas. If the Bwindi gorillas are biologically split from the Virunga gorillas, the world has effectively "lost" half its mountain gorillas and gained another critically endangered gorilla population. Some would argue that the new status might draw even more attention to the gorillas' plight, enhancing their protection. But what if one subspecies thrives while the other declines? If the bloodlines must be kept separate, gorillas from one group could not be recruited into the other, and the weak group might well become extinct. Taxonomists, it seems, could end up killing the mountain gorilla as effectively as poachers do.

At the same time, political instability is a rising threat to the tourist business. For years, Rwandan rebels have occasionally shot at gorilla-watchers in Congo, and in 1994, when genocidal war broke out in Rwanda, most gorilla research and ecotourism in that country and Congo were shut down. Tourists and investigators still felt relatively insulated from the trouble. But as early as last August there were clear signs that they were not. That month a band of Interahamwe rebels kidnapped a group of eight Western tourists who were following gorillas in Congo. Four of those hostages — two New Zealanders and two Swedes — were never heard from again.

Only six months later, the incident at Buhoma brought the danger across the border to Uganda.

When I first heard about the rebel attack I was in Pasadena, California, with my wife and three children, only three days after returning from Africa. I had originally planned to bring the family to my field station, as I had done when working in Tanzania. But the logistical hassles finally convinced my wife and me against it.

In the middle of the night, therefore, when an e-mail message arrived from a colleague in Rwanda, saying "ATTACK ON BUHOMA. MORE DETAILS LATER," I felt a faint trace of relief mixed with my concern. Unfortunately, the details that followed soon dashed that small comfort.

Although the park rangers managed to wound at least four of the attackers, they were quickly overwhelmed. In the next hour the tourism center and several vehicles were set on fire, and the deputy warden, Paul Ross Wagaba, was killed. When the fourteen hostages, nearly all of them tourists, had been rounded up, they were divided into two groups, ordered to take off their shoes, and pushed into the forest barefoot on a forced march back to Congo.

At first I didn't believe that Mitch could be among them. He and I had parted only a few days before in Kampala, the Ugandan capital, and the itinerary he described to me then would not have put him in Buhoma on the morning of the attack. When I called to reassure his parents, however, at their farm in Three Hills, Alberta, Mitch's mother told me that he had called her after I left Uganda. He had changed his plans slightly, she said, and had been in the ecotourism center on that morning after all. There were thirteen tourists in the group, a Canadian official had told her, and one investigator.

A hike through the Ugandan forest can be exhausting even for someone in great shape; for those hostages it must have been excruciating. Mitch's group of six hostages climbed for several hours, up steep forest trails. When they reached the border between Uganda and Congo, they convinced the rebels to express their grievances in a two-page note, composed in French on the spot. "This is a punishment for the Anglo-Saxons who sold us out," the note read, in part. "You are protecting the minority and oppressing the majority." Then Mitch and the others were released. They later learned that the eight hostages in the second group had been brutally murdered.

A week after his ordeal, Mitch flew home to Three Hills. But he is already back in Uganda, working for the International Gorilla Conservation Programme in Kabale. Although I canceled my next research trip, my Ugandan field assistants continue to collect data every day at my camp, and the national park was reopened to tourism only a month after the attack had shut it down. Three tourists a day, on average, have visited the gorillas since, but that is down from twelve or more before the attack. If poaching patrols and basic operations are to continue, the shortfall in income will have to be made up by nongovernmental organizations; regardless, gorilla conservation efforts are likely to suffer.

There is still cause for hope. Two years ago representatives from Congo, Rwanda, and Uganda met for the first time to define shared goals for gorilla conservation. Although their political troubles run deep, their economic incentives for protecting the gorillas run even deeper. Nevertheless, as effective as ecotourism can be in Bwindi, the attack was a lesson in its limitations. Tourist dollars may give local villagers a reason to share in the ethic of protecting gorillas. But money alone can't put out the political fires that so recently spread to Buhoma.

In the past two centuries, the people of Africa have suffered under the same forces that have decimated gorilla populations and destroyed much of the animals' habitat. Myriad governments have been born and died, their countries carved and recarved at the whims of colonial powers and dictators, their people pushed together or shoved apart with no regard for their own ancient ethnic rivalries. Uganda was a British colony; Congo was a Belgian colony; Rwanda was part of German East Africa and then forced into the Belgian League of Nations. All three countries are inherently flammable, and Western aid and Western tourists, in some ways, only add fuel to the blaze.

Whereas my colleagues and I were understandably preoccupied with the Western victims, many Ugandans reacted differently. The deaths of a few tourists at the rebels' hands, they pointed out, provoked nearly as much sympathy and outrage from Westerners as the deaths of a million Rwandans in 1994. Once again, Westerners were demonstrating their neocolonial myopia, showing more concern for white people who happen to be living or traveling in Africa than for the Africans themselves.

The mountain gorillas, for their part, know too little to care: their travels mock the very idea of political nation-states. As long as they remain pawns in the battle zone that is east-central Africa, however, they are in mortal danger, and even the most dedicated conservationists are helpless to protect them. Through science we may try to fix what people have botched through politics and economics. We may declare some species endangered and others healthy, drawing new borders between species or erasing them altogether. But in Africa, the divisions that matter the most are the ones between people, not between animals, and they cut deeper than any border on a map. Only by healing those rifts can we ever hope to save the mountain gorillas.

GARY TAUBES

String Theorists
Find a Rosetta Stone

FROM *Science*

THERE IS A BELIEF, promulgated by Dante, Joseph Campbell, and the makers of major motion pictures, that to get out of a bad situation you must pass through the depths of the abyss. Theoretical physicists have lately taken up this philosophy, although the hell through which they must travel is the guts of black holes — not the kind in the universe at large, but what physicists often refer to, pace Einstein and his thought experiments, as *gedanken* black holes (or, in the words of Princeton University theorist Curt Callan, "*gedanken* black holes to the max").

These hypothetical objects resemble elementary particles more than anything else, and, if real, would be smaller than a hundredth of a quadrillionth of a quadrillionth of a centimeter across. Nonetheless, they have lately taken on the leading role in string theory, physicists' most recent attempt to create a "theory of everything" that unites the forces operating on the microscopic scale of quantum mechanics with the large-scale force of gravity. *Gedanken* black holes have become, in effect, a Rosetta stone, says Andrew Strominger of Harvard University. In the physics of these hypothetical objects, the same phenomena can be found written in the languages of both quantum field theory and general relativity, Einstein's theory of gravity. "These are the two great achievements of twentieth-century physics," says Strominger, "and for the first time we're seeing, at least in some cases, that they are really two sides of the same coin."

If this latest string theory revolution turns out to describe the universe we live in — an enormous if — it will give physicists an unprecedented tool with which to finally develop a quantum theory of gravity. It will allow them to interpret the force of gravity not just according to the rules of general relativity — as the curvature of space-time caused by the presence of matter — but as the result of quantum mechanical fluctuations of the infinitesimal strings out of which, says the theory, all matter is composed. Whether or not the latest work leads to a working theory of everything, it is already responsible for a paradigm shift in how string theorists think about gravity, and it seems poised to provide solutions to some of the most perplexing paradoxes in the field.

The pursuit of a theory of everything and a viable quantum theory of gravity is predicated on a simple fact: Extrapolations from experimental data imply that at a scale of energy known as the Planck scale — some eighteen orders of magnitude beyond what even the most powerful particle accelerators can generate — the gravitational force of the universe at large and the two forces of the microscopic universe, known as the electroweak force and the strong force, would be equally strong and potentially indistinguishable. A theory that describes this unification, casting gravity in the same terms as the electroweak and strong forces, "is what we really need for a complete description of nature," Strominger says.

Reasons to discount string theory as a candidate have always been easy to come by. The theory postulates that the universe is made from tiny, vibrating, stringlike particles, which can be closed loops like rubber bands or open-ended like bits of twine, and multidimensional membranes. Their different modes of vibration, akin to the harmonics on a violin string, would correspond to the different particles and forces in the universe. But this conception is supported by no deep theoretical or geometric insights — it's simply what the equations of the theory happen to describe. Edward Witten of the Institute for Advanced Study in Princeton, New Jersey, the field's impresario, calls the lack of any fundamental principles underlying string theory "the big mystery." Nevertheless, he is sure such principles must exist. "Just as general relativity is based on Einstein's concepts of geometry," he says, "string theory is based on deeper geometric ideas that we haven't yet understood. One facet of this fact is that we can't write down the succinct fundamental equations from which everything else should follow. We've discovered all kinds of equations but not the most fundamental ones."

To complicate matters further, the theory exists in ten or eleven dimensions, six or seven of which are compactified, as physicists call it, in the universe we live in. They are curled up tightly in such infinitesimal spaces that they go unnoticed, leaving four dimensions — three space, one time. All this counterintuitive weirdness aside, the biggest barrier to the acceptance of string theory as a theory of everything is that it so far provides no compelling predictions that can be tested by experiment. The problem is those eighteen orders of magnitude to the Planck scale unification: "The physics is still extremely remote," says Lenny Susskind of Stanford University.

Despite all its drawbacks, however, string theory has been the subject of thousands of papers since 1984, when it emerged as a potential theory of everything, and over four hundred physicists have registered for the latest meeting in Potsdam, Germany, this month. Universities, once hesitant to hire string theorists, have been competing vigorously to get them on campus. Harvard, for example, lured Strominger away from the University of California (UC), Santa Barbara, while Stanford snatched Steve Shenker from Rutgers University, and Princeton snagged Eric and Hermann Verlinde, twins who had been tenured at separate institutions in the Netherlands.

"It is astounding and probably unprecedented that there would be that level of activity for that long in an area which so far has absolutely no tie to experiment," says University of Chicago theorist Jeff Harvey. "The reason we keep on with it is that it seems to lead to new physical insights and beautiful things, wonderful structures. While that may not be proof, it's sufficiently convincing that there's either something to it, or it's got all the best minds in particle theory completely hornswoggled."

Without experiment to guide them, string theory practitioners engage in the theoretical equivalent, testing what happens to their equations when they push the relevant parameters to their limits — when they make the forces between strings extremely strong, for instance, or when they add more or less symmetry to their equations. The more symmetry they add, the more constraints they put on the problems, making them easier to solve, if less realistic.

String theorists talk about tweaking these parameters as though they're playing with the dials on a stereo to see what happens to the music. "Much of the progress we make," says UC Santa Barbara theorist Joe Polchinski, "just comes from taking things we know and trying to push them further, or by looking for puzzles where we can't understand what the physics does when we vary parameters. In that sense it is almost like experimental physics: We don't really know what the theory is. We know a lot of things about it, and we're accumulating facts and trying to put them together into a theory."

The work then proceeds in a manner unique to science. Because practitioners publish their work electronically, through the e-print archives at the Los Alamos National Laboratory in New Mexico, the entire community can read a paper hours after its authors finish typing the last footnote. As a result, no one theorist or even a collaboration does definitive work. Instead, the field progresses like a jazz performance: a few theorists develop a theme, which others quickly take up and elaborate. By the time it's fully developed, a few dozen physicists, working anywhere from Princeton to Bombay to the beaches of Santa Barbara, may have played important parts.

Since 1996 the quantum mechanical properties of black holes have been the dominant theme in this performance, and the field has been playing it with a passion. "Everybody is working on it in one way or another," says Strominger — evidence of either the power of the approach or what Princeton theorist Igor Klebanov calls "a certain amount of herd mentality in the field."

The Second Revolution Continues

The latest series of breakthroughs, assuming they pan out, constitutes the second half of the second revolution in string theory. The first revolution came in 1984, when theorists realized that the theory could conceivably account for all the particles and forces in the universe. The second began a decade later and led directly to the latest progress. Until 1995, string theorists could study the behavior of their equations only in the simplest possible cases, consisting of a few elementary strings with very weak interactions between them. That wasn't adequate for testing the behavior of strings when they interact strongly, under conditions like those in the atomic nucleus or the innards of a black hole. "With strong interactions, where everything is pulling strongly on everything else, you can't use such a simple approximation," says Polchinski. "We had no good tools for understanding what's happening."

In 1995, however, string theorists discovered a wealth of what they call "dualities." These were pairs of equations that allowed them to understand what happens when strings interact strongly, by working instead on "dual" formulations of the relevant equations at weak interactions. "With these dualities," says Polchinski, "we had a whole new, remarkable set of tools to map out how physics changes when interactions become strong. To make an analogy to water, it was like prior to 1995, we knew about steam but nothing else. Now we know about steam and water and ice, and how they change when you change the parameters. It was that kind of conceptual leap."

With their new understanding, the string theorists also found that their equations seemed to describe a slew of potentially fundamental particles — "not just strings," says Harvey, "but membranes or blobs or higher-dimensional widgets." It would take another year to figure out what role these objects played in the theory. In September 1995 Polchinski settled the issue, providing the springboard for the latest progress. Klebanov calls it "the lightning bolt" and says, "all we've been doing since then is milking various applications of his idea."

Polchinski realized that the astonishing new multidimensional objects were all variations on D-branes — objects that he and two students, Jin Dai and Rob Leigh, had identified in 1989 without recognizing their full importance. Some D-branes are stringlike and one-dimensional, while others are surfaces in two, three, or more dimensions. Polchinski, Dai, and Leigh defined them simply as the surfaces on which open strings could end, just as a table leg ends at a table.

Polchinski used dualities to reexamine his D-branes and came up with a comprehensive set of rules for calculating the quantum dynamics of these new objects and understanding their role in the string theory universe. Among the more remarkable properties of Polchinski's D-branes was that their electromagnetic repulsion and their gravitational attraction canceled each other out. As a result, at least on paper and in the imaginations of theorists, they could be stacked on top of each other to create massive objects — objects, as Callan says, "that can be as heavy as you like." For instance, you could wrap one-dimensional stringlike D-branes around one of the tiny compactified dimensions of the string theory universe, or you could wrap multidimensional D-branes around multidimensional compactified spaces, then add more D-branes, inexorably piling on mass. "You make these little Tinkertoy constructions of these wrapped D-branes," says Harvey, "and if you do it in the right way, you get something which at large distances is indistinguishable from a black hole."

In fact, such a Tinkertoy construction, if it existed, would manifest all the properties of a black hole as defined by the rules of general relativity, even though it was constructed purely from the stuff of string theory. It would be so massive as to trap light within its gravitational field; it would have an event horizon, beyond which no light or anything else can escape. And, most critical to what would follow, it would also have a temperature and an entropy, which is a thermodynamic concept that can be thought of as the amount of disorder or randomness in a system. Entropy can also be thought of as the number of different ways you can generate the energy of a system from the combined energy of all its microscopic constituents (atoms, for instance, or molecules — or strings, or even D-branes). Entropy is usually lowest when the temperature of the system is precisely absolute zero. As the temperature rises, the number of different possible states of the system that can generate the energy increases, as does the entropy.

In the mid-1970s, Cambridge University physicists Stephen Hawking and Jacob Bekenstein, now of the Hebrew University of Jerusalem, had demonstrated that the familiar kind of black holes, described by general relativity, must have both a temperature and an entropy, and that they obey a set of laws equivalent to the laws of thermodynamics in gases. In gases, as Ludwig Boltzmann had shown in the nineteenth century, entropy could be derived by counting all the microscopic configurations that molecules in the gas could adopt, which physicists call the microstates of the system. So Hawking and Bekenstein's result implied that a black hole's entropy could be calculated not just by its description from general relativity, known as the Bekenstein-Hawking formula, but by counting microstates. And that, in turn, strongly implied that black holes had a microscopic description, making them a potential bridge between the macroworld and the microworld.

Such a description was beyond any theories of the time. To be meaningful, it would have to satisfy three constraints, says Strominger. "One, it had to include quantum mechanics; two, obviously, it had to include gravity, because black holes are the quintessential gravitational objects. And three, it had to be a theory in which we're able to do the difficult computations of strong interactions, because the forces inside black holes are large. String theory has these first two features: it includes quantum mechanics and gravity. But until 1995, the kinds of things we could calculate were pretty limited."

Polchinski's work on D-branes provided the tools to do the calculations. If a black hole consisted of D-branes stacked together, physicists might be able to convert its microscopic properties into its entropy. Strominger and Harvard theorist Cumrun Vafa looked for a theoretical black hole that they could build out of D-branes, then find its entropy by counting microstates. "There are all kinds of black holes with different numbers of dimensions and different charges and so on," Strominger says. "What we discovered was a black hole in five dimensions. We ended up with this one because it was the only one we could map to a problem we could solve."

To be precise, Strominger and Vafa took the ten dimensions of the string theory universe and compactified them down to five. Then they wrapped five-dimensional and one-dimensional D-branes around the compactified dimensions, ending up with what Strominger calls a "rather complicated bound state of D-branes, contorted and twisted, wrapping around the internal dimensions." Because D-branes are defined as the surfaces on which strings end, strings were also stuck to the D-branes, and these strings, like the D-branes, had excitations running around them. Strominger and Vafa used statistical techniques to count all the possible quantum states of this tangle of strings and D-branes, giving them one measure of the entropy. Then they applied the Bekenstein-Hawking formula, based on general relativity, to find the other. The two agreed exactly.

Impressive as that agreement was, the work still generated skepticism. For all its twists and contortions, the black hole that Strominger and Vafa had constructed, called an extremal black hole, is the simplest of *gedanken* black holes, and its simplicity made it a questionable example. Black holes slowly evaporate through a process known as Hawking radiation. Unlike other black holes, however, extremal black holes carry an electromagnetic charge; as they evaporate, the electromagnetic force eventually cancels out the evaporation and halts the process. That makes extremal black holes relatively easy to work with, says Strominger, "because they're not changing. They're just sitting there."

Theorists worried that what works for tidy extremal black holes might not work for more complicated and more interesting black holes — the "gray bodies," for instance, that are still emitting Hawking radiation. In the eighteen months that followed, says Strominger, various teams tried "to build a more precise dictionary relating these two descriptions of black holes." First, Callan and Juan Maldacena, who was then his graduate student, and independently, Strominger and Gary Horowitz of UC Santa Barbara, constructed and calculated the entropy for what Callan calls "the next more complicated" black hole. They imagined what would happen if their hypothetical tangle of D-branes and strings had vibrations traveling in opposite directions. These waves could collide, annihilate each other, and emit a suitably stringlike particle that would be free to escape the system, taking energy with it. "That coincides to a temperature" that can be calculated, Callan explains. "And then you can reexpress it in terms of the properties of the general relativity black hole that this thing is modeling. And, son of a gun, it gives you exactly the Hawking formulas; you get the right Hawking temperature and the right Hawking radiation rate. These things match beautifully."

Then Klebanov and Steven Gubser, of Princeton, and Amanda Peet, now at UC Santa Barbara, tried to do a similar calculation with four-dimensional D-branes, rather than the tangle of five- and one-dimensional ones. After all, four dimensions — three spatial and one temporal — "is almost our world," says Klebanov. "We did it, and it seemed to almost work." Next, Samir Mathur of the Massachusetts Institute of Technology and Sumit Das of the Tata Institute of Fundamental Research in Bombay, India, calculated how quickly a black hole with some temperature would cool down to its ground state at absolute zero. This time, the two descriptions agreed perfectly.

The next paper, says Callan, was "even more amazing." Strominger and Maldacena teamed up to calculate the dual descriptions of the total energy spectrum of the radiation emitted by a black hole, which is shaped by its gravity as well as temperature. When they calculated this spectrum from the equations of string theory, they got what Callan describes as "some crazy function" — which happened to agree exactly with the result given by general relativity. "Bingo," says Callan. "This is telling you something really uncanny is going on."

Maldacena's Conjecture

By this time, Maldacena, who had moved to Harvard, had begun to put his finger on that uncanny something, the reason why these various descriptions of the black hole and its behavior agreed so well. The quantum theory that described the excitations on D-branes and strings, he speculated, was not only describing the quantum states inside the black hole "but somehow also the geometry of the black hole close to its event horizon" — that is, close to the edge of the black hole. He formulated these thoughts into what string theorists now call Maldacena's conjecture, which states, for certain cases, that a quantum theory with gravity and strings in a given space is completely equivalent to an ordinary quantum system without gravity that lives on the boundary of that space. The conjecture represents the zenith — so far, at least — of revolution number two.

Maldacena took the two ways of looking at a black hole within string theory (one as the quantum mechanical tangle of D-branes and the other as the massive object described by general relativity) and studied what happens to them when the temperature of the black hole approaches absolute zero. On the D-brane side, as energy — and hence mass — dwindles at this low temperature limit, the gravitational force goes to zero, as do the interactions among the strings and the D-branes. What's left is a simple species of quantum field theory called a gauge theory, which is familiar to physicists because, among other reasons, it describes the electroweak and strong forces.

On the general relativity side of the equations, the result of lowering the temperature was even more surprising, says Strominger:

"You start out with a universe with a black hole and strings in it, and then you lower the temperature. As you do that, the space-time of the universe literally gets frozen. Nothing can happen in it except for very, very near the event horizon of the black hole, where there will always be a region hot enough to allow strings to move around freely. The lower the temperature in the space-time, the closer to the horizon that region is forced to be." The space-time of the wider universe is a complex mixture of flat and curved regions, but in this sliver of a region close to the event horizon, it becomes more uniform and symmetrical. The simpler geometry simplifies the string theory that lives in it. "At the end of the day, what's left is a quantum theory of gravity [that] is considerably simpler than the theory we started out with," says Strominger.

This was, in effect, a first glimpse of the ultimate (so far) duality: The complicated D-brane gamische that made up the black hole had reduced to a simple field theory without even strings in it; the general relativity description of the black hole had reduced to a simple quantum theory of gravity. "Before this, it was thought that gravity theories are inherently different from field theories," says Maldacena. "Now we could say that the gravity theory is the same as the field theory."

Everybody in the field seems to interpret this duality a little differently, depending on their own mental images of the universe. Jeff Harvey, for instance, understands it to mean that instead of having two different ways to compute what's going on with the black hole, you now need only one — the microscopic string way. "Maldacena's conjecture says that all the things of ordinary gravity are somehow contained within what happens on these D-branes. It is almost as though gravity is some kind of residual field left over when you treat these other gauge theories in the right way." To Polchinski, the conjecture suggests that D-branes are somehow the atoms from which black holes are made, and gravity is just the combined effect of all these excited strings and D-branes furiously undergoing quantum fluctuations.

The conjecture also had potent implications extending well beyond black holes. In particular, Maldacena's most precise formulation of the conjecture linked the simple string theory near the horizon of a black hole to a particular gauge theory known as a large N gauge theory, which had mystified physicists for twenty-five years.

Gerhard t'Hooft, of the University of Utrecht in the Netherlands, had worked on these theories in the 1970s to understand the behavior of the strong force of the atomic nucleus, which theorists can only precisely calculate at very high energies or temperatures. (*N* stands for the number of colors in the theory; colors are one of the two strong-force equivalents of electromagnetic charge.) He also suggested that such large *N* gauge theories might in fact be string theories, because when physicists drew diagrams describing the interaction of elementary particles in these gauge theories, the diagrams looked a lot like the interactions of strings. T'Hooft did this work well before string theory emerged as a potential theory of everything. "The fantasy that gauge theory might really be a string theory has kicked around for a long time," says Stanford's Steve Shenker.

The catch was, no one had made much progress on the large *N* gauge theories, either. Now, through the circuitous route of black holes and string theory — "a minor miracle," says Klebanov — Maldacena's conjecture suggested a connection between this four-dimensional, large *N* gauge theory and a simple string theory that physicists knew how to solve. By making the large *N* theory tractable, the connection may even be a route to understanding the strong interactions.

Two papers followed Maldacena's, one by Witten and one by Princeton collaborators Klebanov, Gubser, and Alexander Polyakov, who had done crucial work on the four-dimensional gauge theories. These provided specific recipes for what theorists can and cannot calculate when they use a string theory to understand its dual gauge theory. The two papers, says Callan, "very specifically showed what's the rule, what's the recipe, how do you make this connection. And people have been elaborating on it ever since."

So although Maldacena's conjecture has yet to yield any further profound understanding of string theory, string theory is allowing physicists to make progress on theoretical questions that lie much closer to the real world. Whether revolution number two will yield a theory of the universe we live in is still a wide-open question, however. As Strominger says, the work has produced everything a theoretical physicist could want, "except for an experimentally verifiable prediction. It gives us a very precise and explicit relationship between seemingly disparate fields of investigation, a relationship that we can use in some cases to solve problems we've wanted to solve for decades. And it has suggested new ideas that we've previously lacked the imagination to think about."

Once again, string theorists have made enormous progress, but if you ask them where that progress is leading them, they'll still admit that they have no idea. "Part of the goal," says Harvey, "is still to figure out what the hell it all has to do with reality."

Contributors' Notes

Natalie Angier is a science reporter for the *New York Times*. A founding staff member of *Discover*, she went on to become a senior science writer for *Time*, an editor at the women's business magazine, *Savvy*, and a professor at New York University's Graduate Program in Science and Environmental Reporting. She won the Pulitzer Prize for beat reporting in 1991. Her books include *Natural Obsessions* and *The Beauty of the Beastly*, each of which was named a New York Times Notable Book, and *Woman: An Intimate Geography*, which was a finalist for the National Book Award.

An essayist, novelist, and poet, **Wendell Berry** is the author of more than thirty books, including *The Unsettling of America, The Memory of Old Jack, Life Is a Miracle: An Essay Against Modern Superstition*, and *Jayber Crow*. Throughout his career, he has received numerous awards and honors, including an award for writing from the National Institute of Arts and Letters, the Lannan Foundation Award for nonfiction, and the T. S. Eliot Award. He lives and works in his native Kentucky with his wife, Tanya Berry.

Richard Conniff writes about the natural world and about human cultures for *Time, Smithsonian*, the *Atlantic Monthly*, the *New York Times Magazine, National Geographic, Outside, Discover*, and other publications in the United States and abroad. He won the 1997 National Magazine Award for articles in *Smithsonian* excerpted from his book *Spineless Wonders: Strange Tales from the Invertebrate World*. His most recent book, *Every Creeping Thing: True Tales of Faintly Repulsive Wildlife*, was a Book-of-the-Month Club alternate. A former managing editor of *Geo*, Conniff lives in Connecticut.

Paul De Palma is a professor of computer science at Gonzaga University. His interests include the social implications of computing and artificial intelligence. He is currently working on a collection of essays about computing entitled *Dim Sum for the Mind*. He lives in Spokane, Washington, with his wife and daughter.

Helen Epstein holds a Ph.D. in molecular biology and has taught biochemistry at Makerere University in Uganda and conducted AIDS research there. She has written for *Granta, New Scientist,* the *New York Review of Books,* and other magazines and has worked as a consultant for organizations such as the Panos Institute, the London School of Hygiene and Tropical Medicine, and the Rockefeller Foundation.

Anne Fadiman is the editor of the *American Scholar,* a quarterly that has been published since 1932 by the Phi Beta Kappa Society. She is also the author of *Ex Libris: Confessions of a Common Reader* and *The Spirit Catches You and You Fall Down,* which won the 1997 National Book Critics Circle Award for general nonfiction, among other awards. Fadiman's essays and articles have appeared in *Harper's Magazine,* the *New York Times, The New Yorker,* the *Washington Post,* and *Harvard Magazine,* among other publications. While she was a staff writer at *Life,* Fadiman won the 1987 National Magazine Award for reporting for her articles on suicide among the elderly.

Atul Gawande is a surgical resident and a staff writer for *The New Yorker.* He also conducts public-health research at Harvard Medical School. He has served in government on several occasions, most recently as a senior health policy adviser in the Clinton administration. He received his M.D. from Harvard Medical School, an M.A. in politics, philosophy, and economics from Oxford University, and an M.P.H. from the Harvard School of Public Health.

Brian Hayes is a former editor of *Scientific American* and *American Scientist.* He is at work on a book entitled *Infrastructures: A Field Guide to the Industrial Landscape,* to be published in 2001. In 1999 he was journalist in residence at the Mathematical Sciences Research Institute in Berkeley, California. "Clock of Ages" won the 2000 National Magazine Award for essays.

Edward Hoagland is the author of sixteen books, including eleven volumes of essays and nonfiction, among them *African Calliope: A Journey to the Sudan, Notes from the Century Before: A Journal from British Columbia,* and *Tigers and Ice.* His memoir will be published this year.

Judith Hooper is the author of numerous magazine articles and two books, *The Three-Pound Universe* and *Would the Buddha Wear a Walkman?*, and is currently working on a book on evolutionary biology. She lives in western Massachusetts with her husband and son.

Wendy Johnson combines a love of Buddhism (which she has practiced since 1971) with a fierce passion for organic gardening and environmental activism. She writes a regular gardening column for *Tricycle* and is finishing a book on meditation and the culture of organic gardening. Married and the mother of two children, she has lived at Green Gulch Farm Zen Center, a retreat north of San Francisco, for the past twenty-five years.

Ken Lamberton was an award-winning science teacher prior to his incarceration in the late 1980s, after which he began writing. He has published more than a hundred articles in magazines such as *New Mexico Wildlife*, *Arizona Highways*, and *Bird Watcher's Digest*. His first book, *Wilderness and Razor Wire*, was published in 1998.

Peter Matthiessen is the acclaimed author of numerous works of fiction and nonfiction. His accomplishments as a naturalist and explorer have resulted in more than a dozen books on natural history and the environment, including *The Tree Where Man Was Born*, which was nominated for a National Book Award, and *The Snow Leopard*, which won one. His equally distinguished career in fiction has produced a collection of stories and eight novels, among them *At Play in the Fields of the Lord*, which was nominated for a National Book Award, *Far Tortuga*, and the Everglades trilogy, which includes his most recent book, *Bone by Bone*.

Cullen Murphy is the managing editor of the *Atlantic Monthly*, to which he regularly contributes essays and reportage. He also writes the syndicated comic strip *Prince Valiant*, which appears in some 350 newspapers worldwide and which is drawn by his father, John Cullen Murphy. He is the author, with William L. Rathje, of *Rubbish! The Archaeology of Garbage*, and of *Just Curious*, a collection of essays. His most recent book is *The Word According to Eve: Women and the Bible in Ancient Times and Our Own*.

Richard Preston is the author of *The Cobra Event*, a biomedical thriller; *The Hot Zone*, about the Ebola virus; *American Steel*, about a revolutionary steel mill; and *First Light*, about modern astronomy. He is a contributor to *The New Yorker* and has won numerous awards, including the McDermott Award in the Arts from MIT, the American Institute of Physics Award in science writing, and the Overseas Press Club of America Whitman Basso Award for best reporting in any medium on environmental issues. "The Demon in the Freezer" won the 2000 National Magazine Award for public interest writing.

Oliver Sacks was born in London in 1933 and educated in London, Oxford, and California. He is a professor of neurology at the Albert Einstein College of Medicine and the author of seven books, including *The Man Who Mistook His Wife for a Hat* and *The Island of the Colorblind*. The feature films *Awakenings* and *At First Sight* were based on Dr. Sacks's work, and he was the host of the BBC *Mind Traveler* series. He lives in New York, where he swims and raises cycads and ferns.

A native of Memphis, **Hampton Sides** now lives in Santa Fe, New Mexico, where he is a contributing editor for *Outside*. He is the author of *Stomping Grounds*, a collection of nonfiction stories about American subcultures. His work has appeared in the *New York Times Magazine*, *Preservation*, and *DoubleTake*, among other publications, as well as on NPR's *All Things Considered*. His second book, a narrative about survivors of the Bataan Death March, is forthcoming.

Craig B. Stanford is a professor of anthropology and the codirector of the Jane Goodall Research Center at the University of Southern California. He has conducted field studies of chimpanzees and gorillas in Africa and of monkeys in Asia and is the author of seventy scientific publications, including the recent books *Chimpanzee and Red Colobus* and *The Hunting Apes*. He is also the editor of the forthcoming *Meat-Eating and Human Evolution*. He is at work on a book about the biological and cultural foundations of human behavior.

Gary Taubes has written about science, medicine, and health for *Science*, *Discover*, the *Atlantic Monthly*, the *New York Times Magazine*, *Esquire*, *GQ*, and a host of other publications. He has won numerous awards for his reporting, including the National Association of Science Writers Science-in-Society Journalism Award in both 1996 and 1999. His most recent book, *Bad Science: The Short Life and Weird Times of Cold Fusion*, was a New York Times Notable Book and a finalist for the Los Angeles Times Book Award.

Other Notable Science and Nature Writing of 1999

SELECTED BY BURKHARD BILGER

DAVID GUTERSON
  The Kingdom of Apples. *Harper's Magazine*, October.

JOHN HALES
  Hiking In, and Hiking Out. *Hudson Review*, Spring.
STEPHEN S. HALL
  Fear Itself. *New York Times Magazine*, February 28.
  Journey to the Center of My Mind. *New York Times Magazine*, June 6.
FRANCIS HALZEN
  Antarctic Dreams. *The Sciences*, March/April.
ROALD HOFFMANN
  A Really Moving Story. *American Scientist*, January/February.

GEORGE JOHNSON
  Terra Incognita. *New York Times Magazine*, October 17.

ROBERT KAPLAN
  Mayan Mathematics: The Dark Side of Zero. *American Scholar*, Summer.
TED KERASOTE
  Tracking in the Snow. *Audubon*, January/February.
GEORGINA KLEEGE
  On the Borders of the Wild. *Raritan*, Spring.
TRAVIS W. KNOWLES
  Twilight on Bald Mountain. *The Sciences*, March/April.
KENNETH S. KOSIK
  The Fortune Teller. *The Sciences*, July/August.
KATE KRAUTKRAMER
  Walking. *High Plains Literary Review*, Fall/Winter.
ROBERT KUNZIG
  A Tale of Two Archaeologists. *Discover*, May.

WILLIAM LANGEWIESCHE
  Eden: A Gated Community. *Atlantic Monthly*, June.
ALAN LIGHTMAN
  A Cataclysm of Thought. *Atlantic Monthly*, January.
STEPHEN LYONS
  Enough Nature Writing Already! *High Country News*, May 10.

CHARLES C. MANN
  Biotech Goes Wild. *Technology Review*, July/August.
JEFFREY M. MASTER
  Hunting Hugo. *Weatherwise*, September/October.
JOHN McPHEE
  Farewell to the Nineteenth Century. *The New Yorker*, September 27.
OLIVER MORTON
  For the Love of Mars. *Discover*, February.